Flexible Thermoelectric Polymers and Systems

Flexible Thermoelectric Polymers and Systems

Edited by

Jianyong Ouyang
National University of Singapore, Singapore

WILEY

This edition first published 2022

Registered Office
John Wiley & Sons, Inc., 111 River Street, Hoboken, NJ 07030, USA

Editorial Office
111 River Street, Hoboken, NJ 07030, USA

For details of our global editorial offices, customer services, and more information about Wiley products visit us at www.wiley.com.

Wiley also publishes its books in a variety of electronic formats and by print-on-demand. Some content that appears in standard print versions of this book may not be available in other formats.

Library of Congress Cataloging-in-Publication Data

Names: Ouyang, Jianyong, author.
Title: Flexible thermoelectric polymers and systems / edited by Jianyong Ouyang, National University of Singapore.
Description: 1st edition. | Hoboken, NJ : Wiley, 2022. | Includes index.
Identifiers: LCCN 2021033949 (print) | LCCN 2021033950 (ebook) | ISBN 9781119550709 (hardback) | ISBN 9781119550723 (ebook) | ISBN 9781119550686 (epdf) | ISBN 9781119550631 (epub)
Subjects: LCSH: Thermoelectric apparatus and appliances–Materials. | Thermoresponsive polymers. | Flexible electronics.
Classification: LCC TK2950 .O987 2022 (print) | LCC TK2950 (ebook) | DDC 621.31/243–dc23
LC record available at https://lccn.loc.gov/2021033949
LC ebook record available at https://lccn.loc.gov/2021033950

Cover Design: Wiley
Cover Image: © Ajwad Creative/Getty

Set in 9.5/12.5pt STIXTwoText by Straive, Pondicherry, India

MIX
Paper from responsible sources
FSC
www.fsc.org
FSC® C013604

10 9 8 7 6 5 4 3 2 1

Contents

List of Contributors

Shanshan Chen
School of Energy & Power Engineering
State Key Laboratory of Power
Transmission Equipment & System
Security and New Technology
Chongqing University
Chongqing
P. R. China

Xinyi Chen
Key Laboratory of Textile Science &
Technology of Ministry of Education
College of Textiles
Donghua University
Shanghai
China

Hanlin Cheng
Department of Materials Science
and Engineering
National University of Singapore
Singapore
Singapore

Ming Hui Chua
Institute of Materials Research and
Engineering (IMRE)
Agency for Science, Research and
Technology (A*STAR)
Singapore
Singapore

Minzhi Du
Key Laboratory of Textile
Science & Technology of Ministry
of Education
College of Textiles
Donghua University
Shanghai
China

Zeng Fan
School of Physics
Dalian University of Technology
Dalian
P. R. China

Xue Han
Key Laboratory of Textile Science
& Technology of Ministry of
Education
College of Textiles
Donghua University
Shanghai
China

Lin Hu
China-Australia Institute
for Advanced Materials and
Manufacturing
Jiaxing University
Jiaxing
P. R. China

Fengxing Jiang
Flexible Electronics Innovation Institute
Jiangxi Science and Technology Normal University
Nanchang
P. R. China
Department of Physics
Jiangxi Science and Technology Normal University
Nanchang
P. R. China

Yuanyuan Jing
Key Laboratory of Textile Science & Technology of Ministry of Education
College of Textiles
Donghua University
Shanghai
China

Meng Li
MOE Key Laboratory of Low-grade Energy Utilization Technologies and Systems
CQU-NUS Renewable Energy Materials & Devices Joint Laboratory
School of Energy & Power Engineering
Chongqing University
Chongqing
P. R. China

Zaifang Li
China-Australia Institute for Advanced Materials and Manufacturing
Jiaxing University
Jiaxing
P. R. China

Peipei Liu
Department of Physics
Jiangxi Science and Technology Normal University
Nanchang
P. R. China

Congcong Liu
Flexible Electronics Innovation Institute
Jiangxi Science and Technology Normal University
Nanchang
P. R. China

Yang Liu
MOE Key Laboratory of Low-grade Energy Utilization Technologies and Systems
CQU-NUS Renewable Energy Materials & Devices Joint Laboratory
School of Energy & Power Engineering
Chongqing University
Chongqing
P. R. China

Baoyang Lu
Flexible Electronics Innovation Institute
Jiangxi Science and Technology Normal University
Nanchang
P. R. China

Chunhong Lu
Key Laboratory of Textile Science & Technology of Ministry of Education
College of Textiles
Donghua University
Shanghai
China

Jianyong Ouyang
Department of Materials Science and Engineering
National University of Singapore
Singapore
Singapore

Lujun Pan
School of Physics
Dalian University of Technology
Dalian
P. R. China

Yue Shu
MOE Key Laboratory of Low-grade Energy Utilization Technologies and Systems
CQU-NUS Renewable Energy Materials & Devices Joint Laboratory
School of Energy & Power Engineering
Chongqing University
Chongqing
P. R. China

Kuan Sun
MOE Key Laboratory of Low-grade Energy Utilization Technologies and Systems
CQU-NUS Renewable Energy Materials & Devices Joint Laboratory
School of Energy & Power Engineering
Chongqing University
Chongqing
P. R. China

Ady Suwardi
Institute of Materials Research and Engineering (IMRE)
Agency for Science, Research and Technology (A*STAR)
Singapore
Singapore

Zhenghong Xiong
MOE Key Laboratory of Low-grade Energy Utilization Technologies and Systems
CQU-NUS Renewable Energy Materials & Devices Joint Laboratory
School of Energy & Power Engineering
Chongqing University
Chongqing
P. R. China

Jianwei Xu
Institute of Materials Research and Engineering (IMRE)
Agency for Science, Research and Technology (A*STAR)
Singapore
Singapore

Jingkun Xu
Flexible Electronics Innovation Institute
Jiangxi Science and Technology Normal University
Nanchang
P. R. China

Yuanyuan Zheng
Key Laboratory of Textile Science & Technology of Ministry of Education
College of Textiles, Donghua University
Shanghai
China

Yujie Zheng
School of Energy & Power Engineering
State Key Laboratory of Power Transmission Equipment & System Security and New Technology
Chongqing University
Chongqing
P. R. China

Fengling Zhang
Department of Physics
Chemistry and Biology
Linköping University
Linköping, Sweden

Kun Zhang
Key Laboratory of Textile Science & Technology of Ministry of Education
College of Textiles
Donghua University
Shanghai
China

Yinhua Zhou
Wuhan National Laboratory for Optoelectronics, and School of Optical and Electronic Information
Huazhong University of Science and Technology
Wuhan
Hubei
China

Yongli Zhou
MOE Key Laboratory of Low-grade Energy Utilization Technologies and Systems
CQU-NUS Renewable Energy Materials & Devices Joint Laboratory
School of Energy & Power Engineering
Chongqing University
Chongqing
P. R. China

Preface

Thermoelectric materials are interesting because they can be used to directly convert heat into electricity. Some conductive materials can exhibit a notable voltage under temperature gradient, that is, the Seebeck effect. The thermovoltage with respect to the temperature is called Seebeck voltage. The Seebeck voltage can be positive or negative depending on the accumulation of the charge carrier(s) at the two ends under temperature gradient. A *p*-type thermoelectric material exhibits a positive Seebeck voltage, while an *n*-type thermoelectric material has a negative Seebeck voltage. High Seebeck coefficient has been observed on semiconductors and semimetals. The value is related to the relative position of the Fermi level from the top of the valence band or the bottom of the conduction band or simply the doping level since a change in the doping level can shift the Fermi level. Apart from the Seebeck coefficient, a high electrical conductivity is also required for efficient thermoelectric conversion. However, a good thermoelectric material should have a low thermal conductivity so that more heat can be utilized for power generation. Thermoelectric materials are used as the active materials of thermoelectric generators. They can be used in thermoelectric cooling systems in terms of the Peltier effect as well.

The conventional thermoelectric materials are inorganic semiconductors or semimetals such as Bi_2Te_3 and its derivatives or analogues. These materials can exhibit high Seebeck coefficient and high electrical conductivity, but they also have a problem of high thermal conductivity. Hence, great effort has been made on lowering the thermal conductivity of inorganic thermoelectric materials besides on the improvement in the Seebeck coefficient and/or electrical conductivity. Consider for practical application, inorganic thermoelectric materials are usually brittle, and thus they are not suitable for flexible thermoelectric systems.

Polymers emerge as the next-generation thermoelectric materials mainly due to their high mechanical flexibility. Flexible thermoelectric materials can enable the realization of flexible or even portable/wearable thermoelectric generators or coolers. They are significant because of the ubiquitous heat on earth and the convenient heat collection related to the mechanical flexibility. For example, a wearable thermoelectric generator can harvest heat from human body and provide electricity to other wearable or implanted devices that are used for communication or healthcare. Polymers have much lower thermal conductivity than their inorganic counterparts, but their Seebeck coefficient is remarkably inferior to the latter. Therefore, the research focus on developing high-performance thermoelectric polymers is very different from developing high-performance inorganic thermoelectric materials.

Thermoelectric polymers are intrinsically conducting polymers in the doped state. They have conjugated backbone, and their charge carriers are produced through the chemical oxidation or reduction. Although intrinsically conducting polymers have been studied for more than 40 years since the discovery of conducting polyacetylene in 1970s, their thermoelectric properties only recently gained considerable attention. Hence, some researchers in the area of organic/polymer electronics may not be very familiar with the knowledge of thermoelectrics. The first chapter of this book introduces the fundamental knowledge of thermoelectric materials, thermoelectric effects, thermoelectric generators, and Peltier coolers. Some fundament knowledge of thermoelectric polymers is notably different from that of inorganic thermoelectric materials. Chapter 2 provides an overall review on the thermoelectric polymers. The thermoelectric properties of polymers depend not only on their chemical structure but also on their doping level and morphology. The most popular thermoelectric polymers are the poly(3,4-ethylenedioxythiophene) (PEDOT) family, and they are reviewed in Chapter 3. Although intrinsically conducting polymers can be prepared through chemical synthesis, electrochemical polymerization is also popular for the preparation of intrinsically conducting polymers. The Chapter 4 reviews the thermoelectric polymers by electrochemical synthesis. Apart from electronic conductors including semiconductors, semimetals and intrinsically conducting polymers, ionic conductors recently attracted great attention owing to their high thermovoltage that is usually higher than that of electronic conductors by two to three orders by magnitude. In addition, they can affect the thermoelectric properties of polymers as well. This is covered in Chapter 5. Chapter 6 provides a comprehensive review on the thermoelectric polymer composites. The nanofillers in the composites can greatly improve the thermoelectric properties of polymers.

The last chapter introduces the thermoelectric properties of low-dimensional thermoelectric materials including zero-dimensional nanoparticles, one-dimensional nanowire or nanotubes, and two-dimensional materials. These low-dimensional materials can exhibit interesting thermoelectric materials, and they are often used as the nanofillers in thermoelectric polymer composites for practical application.

I am grateful to all the authors for their contributions to this book particularly in this special time of the COVID-19 pandemic. I also would like to thank my wife and my two sons for their support.

Jianyong Ouyang

1

Fundamental Knowledge on Thermoelectric Materials

Jianyong Ouyang and Hanlin Cheng

Department of Materials Science and Engineering, National University of Singapore, Singapore, Singapore

1.1 Properties of Thermoelectric Materials

Thermoelectric materials are electronic materials that have charge carrier distributions sensitive to temperature and produce a voltage at the presence of a temperature gradient. In terms of the electrical conductivity, electronic materials can be classified into conductors, semiconductors, and insulators. Conductors like metals usually have a conductivity of $>10^2\ \mathrm{S\,cm^{-1}}$, while insulators like plastics usually have a conductivity of below $10^{-7}\ \mathrm{S\,cm^{-1}}$. The conductivity of semiconductors is between those of the conductors and insulators. The conductivity of a semiconductor can be dramatically increased by doping. The different electrical conductivities of these materials arise from their different electronic band structures. As shown in Figure 1.1, conductors have a partially filled band, and the maximum electron energy at the absolute 0 K is the Fermi energy (E_F). There is a bandgap between the conduction band and valence band for semiconductors. For a semiconductor, the valence band is fully occupied while the conduction band is completely empty at 0 K in dark. The Fermi level is between the top of the valence band and the bottom of the conduction band for an intrinsic or lightly doped semiconductor. Thermal excitation or light exposure can induce electron transition from the valence band to the conduction band. This leads to increase in the electrical conductivity. The bandgap of insulators is much larger than semiconductors. Insulators always have very low conductivity because

Flexible Thermoelectric Polymers and Systems, First Edition. Edited by Jianyong Ouyang.

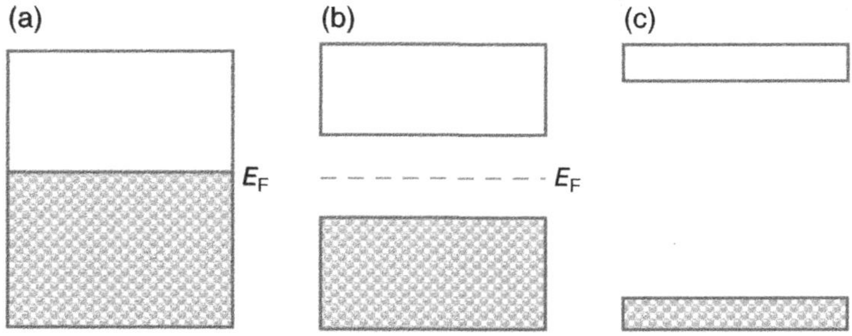

Figure 1.1 Band structures of (a) conductors, (b) semiconductors, and (c) insulators.

temperature or light cannot usually excite the electrons from the valence band to the conduction band.

In terms of the energy band structure, the conventional conductors are metals or heavily doped semiconductors that can have the Fermi level higher than the minimum of the conduction band or lower than the maximum of the valence band. Some charge-transfer organic salts, intrinsically conducting polymers, multi-walled carbon nanotubes (MWCNTs), single-walled carbon nanotubes (SWCNTs) with certain chirality, graphene, and MXenes can have partially filled band structure, and they are thus conductors. The conventional semiconductors include Si, Ge, and compound semiconductors. Conjugated organic compounds and conjugated polymers in neutral state are considered as organic semiconductors. Single-walled carbon nanotubes can be semiconductive or metallic depending on the chirality.

Apart from the electrical conductivity, the Seebeck coefficient and thermal conductivity are important parameters for thermoelectric materials. The thermoelectric performance of a material is usually characterized by the dimensionless figure of merit (ZT),

$$ZT = \frac{S^2 \sigma T}{\kappa},$$

where S is the Seebeck coefficient, σ is the electrical conductivity, T is the absolute temperature, and κ is the thermal conductivity. $S^2\sigma$ is called the power factor. To have a high ZT value, the thermoelectric material should have a high power factor and a low thermal conductivity. Because thermoelectric polymers or organic molecules usually have very low thermal conductivity, the power factor is the important parameter for their thermoelectric performance.

Although metals can have high electrical conductivity, their Seebeck coefficient is usually quite low. Intrinsic semiconductors can have high

Seebeck coefficient but low electrical conductivity. Hence, both of them are not good thermoelectric materials. The conventional thermoelectric materials are inorganic semiconductors or semimetals such as BiTe. They can have high Seebeck coefficient and high electrical conductivity. But their high thermal conductivity is a big concern. Thus, effort in improving the thermoelectric performance of inorganic materials has been focused on lowering the thermal conductivity. Because of low intrinsic thermal conductivity, high mechanical flexibility, and low fabrication cost, organic thermoelectric materials have been gaining increased attention [1–4]. The organic thermoelectric materials are usually conducting polymers or doped conjugated organic molecules. Figure 1.2 presents the chemical structure of some representative conducting polymers. The Seebeck coefficient of organic materials is usually lower than their inorganic counterparts by about one to two orders of magnitude. To achieve high thermoelectric performance, it is important to improve the Seebeck coefficient and/or the electrical conductivity of organic thermoelectric materials.

To develop materials with high thermoelectric properties, good understanding in the Seebeck coefficient, electrical conductivity, and thermal conductivity is needed. Although the knowledge acquired from the study of inorganic thermoelectric materials is often used, it is not always true for organic thermoelectric materials.

1.1.1 Thermoelectric Effect

There are three thermoelectric effects, that is, Seebeck effect, Peltier effect, and Thomson effect. Correspondingly, there are Seebeck coefficient, Peltier coefficient, and Thomson coefficient. The Seebeck and Peltier coefficients are usually used for the thermoelectric generation and cooling, respectively. The Thomson coefficient is used because of the temperature dependence of the Seebeck coefficient. The three coefficients are related, and the Seebeck coefficient is the most popular term used in literature.

1.1.2 Seebeck Effect

Thermoelectric materials are usually referred to electronic materials with high Seebeck coefficient. When there is a temperature gradient between the two ends of a thermoelectric material, a voltage will be generated between the two ends. This effect was first discovered by Alessandro Volta in 1794 and independently rediscovered by Thomas J. Seebeck in 1821. The Seebeck coefficient (S) is determined by the voltage (ΔV) and the temperature gradient (ΔT),

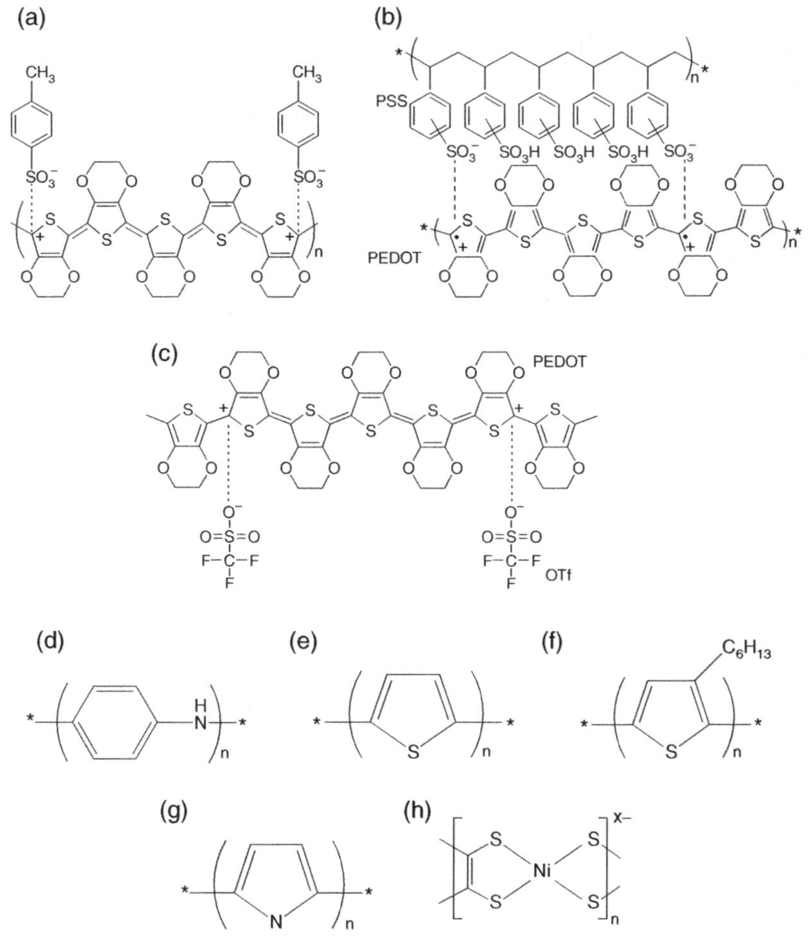

Figure 1.2 Chemical structures of some representative conducting polymers: (a) poly(3,4-ethylenedioxythiophene):tosylate (PEDOT:Tos), (b) poly(3,4-ethylenedioxythiophene): polystyrene sulfonate (PEDOT:PSS), (c) poly(3,4-ethylenedioxythiophene): trifluoromethanesulfonate (PEDOT:OTf), (d) polyaniline (PANI), (e) polythiophene (PTh), (f) poly(3-hexylthipohene) (P3HT), (g) polypyrrole (PPy), and (h) poly(nickel-ethylenetetrathiolate) [poly(Ni-ett)].

$$S = \frac{\Delta V}{\Delta T}. \tag{1.1}$$

The Seebeck effect can be understood in terms of the temperature effect on the charge carrier distribution. For example, the electrons of a metal follow the Fermi–Dirac distribution,

$$f(E) = \frac{1}{1 + \exp\left(\frac{E - E_F}{k_B T}\right)}, \tag{1.2}$$

where E is the electron energy and k_B the Boltzmann constant. The electron distribution near the Fermi level is affected by temperature. The $f(E)$ value at E_F is 1 at 0 K and less than one at a temperature above 0 K. Consider two different temperatures, T_H and T_C, at the two ends of a metal, the Fermi–Dirac distributions at the two temperatures are shown in Figure 1.3. When the temperature (T_C) at the cold side is 0 K, the states above the Fermi level is completely empty, while all the states below the Fermi level is fully occupied. At the hot side, the temperature (T_H) is greater than 0 K. Thus, in terms of the Fermi–Dirac distribution, some states at the energy above E_F is occupied, while some states below the Fermi level are vacant. As a result, electrons with the energy higher than E_F will diffuse from the hot side to the cold side, while the electrons with the energy lower than E_F will migrate in the opposite direction. But the electron diffusions along the two directions are not equivalent. At equilibrium, some electrons accumulate at the hot or cold side, this leads to an internal electric field and a voltage between the two sides. If the electrons accumulate at the cold side, the potential of the cold side will be lower than the hot side. This will lead to a negative Seebeck coefficient, and the material is an n-type thermoelectric material. Instead, if

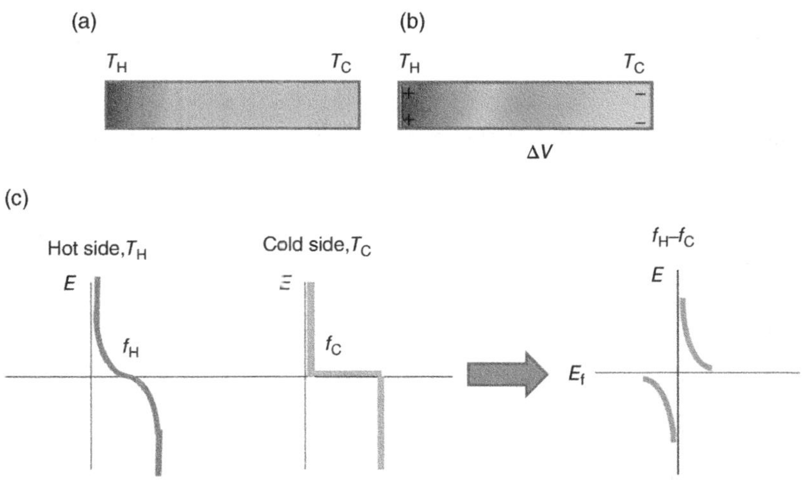

Figure 1.3 (a) A metal with different temperatures at the two sides. (b) Voltage between the two sides induced by temperature gradient. (c) Fermi–Dirac distributions at two different temperatures of T_H and T_C and their difference.

the electrons accumulate at the hot side, the cold side has a potential higher than the hot side. The Seebeck coefficient is positive, and it is a *p*-type material.

By using the Fermi gas model, the Seebeck coefficient of metals is

$$S \approx -\frac{\pi^2 k_B}{3e}\frac{T}{T_F}, \tag{1.3}$$

where T_F is the Fermi temperature. This is the Mott formula in metals. In terms of this formula, metals usually have small Seebeck coefficient because of their large T_F value of 10^4–10^5 K. The absolute Seebeck coefficient values of metals are usually less than 10 μV K^{-1} (Table 1.1).

The Seebeck coefficient of semiconductors depends on the doping level. For nondegenerated semiconductors, Eqs. (1.4) and (1.5) are the Mott formulae for the Seebeck coefficients in the conduction band and valence band, respectively,

$$S_c = -\frac{k_B}{e}\left(\frac{E_c - E_F}{k_B T} + a_c + 1\right), \tag{1.4}$$

$$S_v = \frac{k_B}{e}\left(\frac{E_F - E_v}{k_B T} + a_v + 1\right), \tag{1.5}$$

where E_c is the minimum energy of the conduction band, E_v is the maximum energy of the valence band, and a_c and a_v are constants and they depend on materials. In terms of the Mott formulae in semiconductors, the

Table 1.1 Seebeck coefficient of some metals, semiconductors, flexible materials at room temperature.

Metal	S (μVK^{-1})	Semiconductor	S (μVK^{-1})	Flexible materials	S (μVK^{-1})
Sb	+47	Se	+900	PEDOT:PSS	+14–20
Mo	+10	Te	+500	PEDOT:Tso	+30
Ag	+6.5	Si	+440	PA	+10
Cu	+6.5	Ge	+300	PPy	+10
Hg	+0.6	Sb_2Te_3	+185	PANi	+10–20
Pt	+0.0	$Pb_3Ge_{39}Se_{58}$	+1670	Poly(Ni-ett)	−125
Ni	−15	$Pb_{15}Ge_{37}Se_{58}$	−1990	graphene	−10
Bi	−72	Bi_2Te_3	−230	Carbon nanotube	+20

Seebeck coefficient has the maximum value when a semiconductor is lightly doped.

For degenerated semiconductors, the Seebeck coefficient is given by

$$S = -\frac{\pi^2 k_B T}{3e}\left[\frac{d\ln\sigma(E)}{dE}\right]_{E=E_F}, \tag{1.6}$$

$$\sigma(E) = en(E)\mu(E). \tag{1.7}$$

where $n(e)$ and $\mu(E)$ are the charge carrier density and mobility at energy E. Hence, a semiconductor with a large variation in the carrier density near the Fermi level can have a high Seebeck coefficient. This can be achieved by changing the band structure. This way to increase the Seebeck coefficient is called "band structure engineering". In addition, the charge carrier density $n(E)$ is related to the density of states (DOS), $g(E)$,

$$n(E) = g(E)f(E). \tag{1.8}$$

Thus, the Seebeck coefficient is related to the density of states near E_F. Controlling the density of states can vary the Seebeck coefficient, and this is called "DOS engineering".

Another important formula is the dependence of the Seebeck coefficient on the difference between the average energy (E_J) of the transporting electrons and the E_F,

$$S = -\frac{E_J - E_F}{qT}, \tag{1.9}$$

where q is the elementary charge. A simple picture of an n-type semiconductor is shown in Figure 1.4. The charge carriers are the electrons in the conduction band. Their average energy is E_J. For n-type semiconductors, $E_J - E_F > 0$, the Seebeck coefficient is negative. In contrast, $E_J - E_F < 0$ for p-type semiconductors, the Seebeck coefficient is positive. When the $|E_J - E_F|$ value is large, the material will have a high Seebeck coefficient. Increasing the doping level of an n-type semiconductor can shift the Fermi level upward to the conduction band and thus lower the $|E_J - E_F|$ value. Similar argument can be applied for p-type semiconductors. Hence, the Seebeck coefficient decreases with the increase in the doping level.

Although increasing the doping level can decrease the Seebeck coefficient, it increases the electrical conductivity. The power factor ($S^2\sigma$) is dependent on the Seebeck coefficient and the electrical conductivity (σ). As

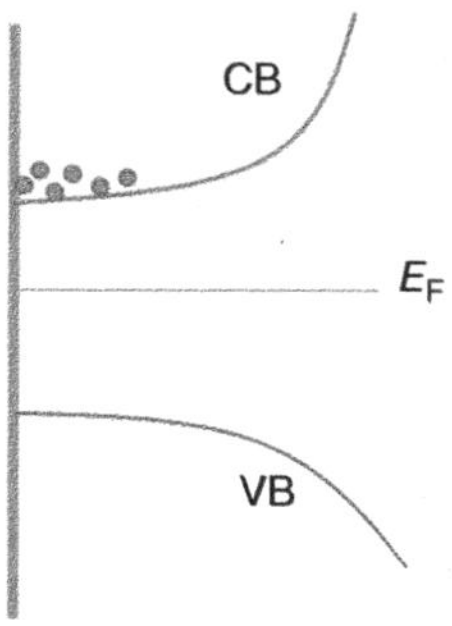

Figure 1.4 Band structure of an *n*-type semiconductor. VB and CB are for the valence band and conduction band, respectively. The dots at the CB stand for electrons.

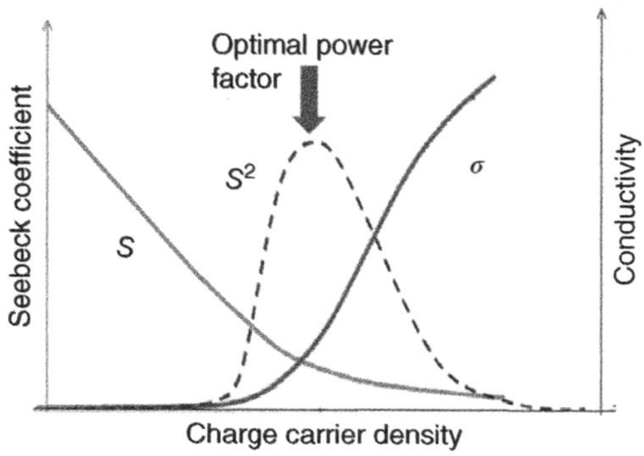

Figure 1.5 Dependence of the Seebeck coefficient, electrical conductivity, and power factor on the charge carrier concentration.

shown in Figure 1.5, there is an optimal power factor in terms of the charge carrier density.

The dependence of the power factor on the charge carrier density is valid for conducting polymers as well. The Seebeck coefficient and electrical conductivity of conducting polymers strongly depend on the doping level. An undoped conducting polymer has a conductivity in the insulator range. The increase in the doping level can decrease the Seebeck coefficient while increase the electrical conductivity. Thus, doping engineering is an effective way to find the optimal power factor of conducting polymers as well. Chemical and electrochemical de-dopings are popular methods to find the optimal doping level of conducting polymers in terms of the power factor. For example, Crispin et al. chemically de-doped PEDOT:TsO with tetrakis(dimethylamino) ethylene (TDAE) and found the optimal doping level of ~0.22 (Figure 1.6) [5].

The Seebeck coefficient of conducting polymers is also related to the doping form. There are two types of doping forms for *p*-type conducting polymers [6, 7]. Taking away electrons from the conjugated backbone is the oxidative doping. Oxidative doping is the main doping form for conjugated polymers such as polyacetylene and polythiophene. The oxidative doping

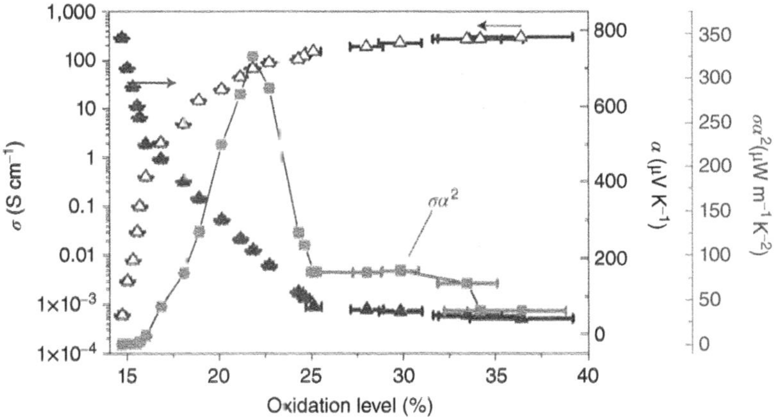

Figure 1.6 Dependences of the electrical conductivity, Seebeck coefficient, and power factor of PEDOT:Tso on the doping level. Oxidation level is used for the doping level, and α s used for the Seebeck coefficient. *Source:* Bubnova et al. [5]. © Springer Nature.

can be de-doped by reducing chemical agents. Another doping form is the protonic acid doping. It is the main doping form for polyaniline because polyaniline has ammine group [8]. Protonic acid doping was observed for polypyrrole and polythiophene as well [6, 7]. The protonic acid doping can be de-doped by base. Yao et al. found that de-doping the protonic acid doping by base is more effective in increasing the Seebeck coefficient than de-doping the oxidative doping [3]. As shown in Figure 1.7, poly(3,4-ethylenedioxythiophene):trifluoromethanesulfonate (PEDOT:OTf) can be de-doped with reducing agents including glucose and ascorbic acid or NaOH. When PEDOT:PSS:OTf is de-doped to the same doping level, the conductivity using glucose or ascorbic acid is similar with NaOH. But the latter can give rise to a higher Seebeck coefficient and thus the power factor than the former. This difference is ascribed to the different electronic structures by oxidative doping and protonic acid doping. The protonic acid doping takes place by attaching a proton to an α or β site of the thiophene ring of PEDOT. This distorts the conjugated structure and leads to more electrons near the Fermi level. Instead, the oxidative doping does not affect the conjugated structure. Thus, the removal of the protonic acid doping can more effectively increase the mean energy (E_J) of the charge carriers and thus increase the Seebeck coefficient more remarkably.

The dependence of the Seebeck coefficient on the $|E_J - E_F|$ values is often used to understand the energy filtering. Energy filtering is often observed when an electronic material like nano-fillers is mixed into a conducting

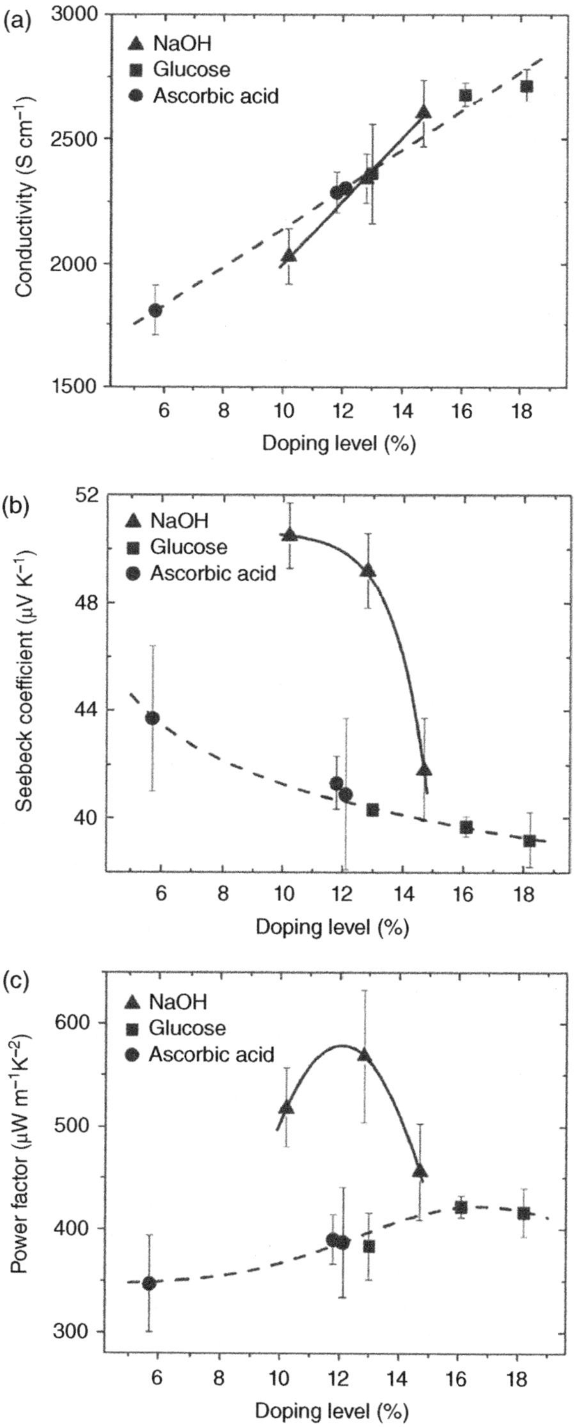

Figure 1.7 (a) Conductivity, (b) Seebeck coefficient, and (c) power factor of PEDOT:OTf films as a function of doping levels. NaOH, glucose, and ascorbic acid were used to de-dope PEDOT:OTf. *Source:* Yao et al. [7]. © Royal Society of Chemistry.

polymer with different Fermi levels [9–11]. The different Fermi levels of the matrix and nano-fillers can induce the charge transfer and an internal electric field at the interface between them. This internal electric field at the interface can block the accumulation of the charge carriers with low energy and thus increase the mean energy (E_J) of the accumulated charge carriers under temperature gradient. Apart from composites of conducting polymers, energy filtering was reported when a certain material is coated on a conducting polymer film [12–14].

1.1.3 Peltier Effect

Electrical current is generated at the presence of temperature gradient in terms of the Seebeck effect. There is a reverse process to take away heat by applying an electrical current to a thermoelectric material. This is called the Peltier effect. When a charge carrier transports from the cold side to the hot side, it will bring heat from the cold side to the hot side. Thus, this can lower the temperature of the cold side.

The Peltier coefficient (Π) is related to the Seebeck coefficient by

$$\Pi = ST. \tag{1.10}$$

In terms of this relationship, a material with a high Seebeck coefficient also has a high Peltier coefficient. Apart from the heat transfer by the Peltier effect, electrical current will generate Joule heat as well. Thus, the conductivity of the thermoelectric materials is also an important parameter for the Peltier cooling.

1.1.4 Thomson Effect

The Seebeck coefficient of materials is dependent on the temperature. A spatial gradient in temperature can generate a gradient in the Seebeck coefficient. When an electrical current flows through a material with temperature gradient, continuous Peltier effect can take place. This is called Thomson effect.

When a current density (J) transports through a homogeneous conductor, the heat generation rate per unit volume will be generated in terms of the Thomson effect,

$$\dot{q} = -K_{\mathrm{Th}} J \frac{dT}{dx}, \tag{1.11}$$

where K_{Th} is the Thomson coefficient and $\mathrm{d}T/\mathrm{d}x$ is the temperature gradient. The Thomson coefficient is related to the Seebeck coefficient by,

$$K_{\mathrm{Th}} = T\frac{dS}{dT}. \tag{1.12}$$

1.1.5 Electrical Conductivity

In terms of the classic theory, the electrical conductivity of electronic materials depends on the charge carrier density (n) and charge carrier mobility (μ),

$$\sigma = ne\mu, \tag{1.13}$$

where e is the elementary charge of an electron.

1.1.5.1 Charge Carrier Density

Metals have high charge carrier density and thus high electrical conductivity. Their charge carrier density depends on the density and valence electrons of the metal atoms. Intrinsic semiconductors have very low charge carrier density, and charge carriers can be generated by temperature excitation or light-induced excitation. The conductivity of a semiconductors can be increased by several orders of magnitude by doping. The electrical conductivity of a semiconductor depends on the electrons in the conduction band and holes in the valence band,

$$\sigma = ne\mu_{\mathrm{e}} + pe\mu_{\mathrm{p}}, \tag{1.14}$$

where n and p are the densities of the electrons in the conduction band and holes in the valence band, respectively, and μ_{e} and μ_{p} are the electron and hole mobilities, respectively.

Figure 1.8 shows the band structure of intrinsic, p-type, and n-type semiconductors. For example, when Si is doped with a pentavalent element like P or As, electrons are generated in the conduction band. It thus becomes an

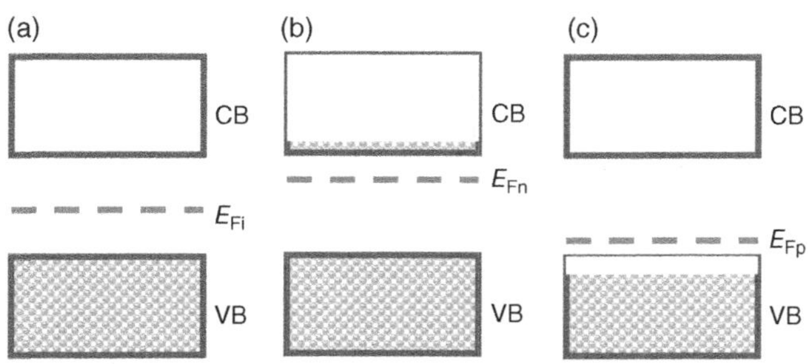

Figure 1.8 Band structures of (a) intrinsic, (b) n-type, and (c) p-type semiconductors.

n-type extrinsic semiconductor. Instead, when the doping element is a trivalent element like B, Al, or Ga, holes are generated in the valence band. The semiconductor becomes a *p*-type extrinsic semiconductor. Although doping can increase the charge carrier density, it also shifts the Fermi level. The *n*-type doping shifts the Fermi level upward to the conduction band, while *p*-doping shifts the Fermi level downward to the valence band. Because the Seebeck coefficient of a lightly doped semiconductor is related to the energy difference between the bottom of the conduction band and the Fermi level for a *n*-type semiconductor and between the top of the valence band and the Fermi level for a *p*-type semiconductor. Therefore, increasing the doping level can increase the electrical conductivity but lower the Seebeck coefficient.

The conductivity of intrinsically conducting polymers can be dramatically increased by doping as well. But the nature of the doping for conducting polymers is fundamentally different from that of inorganic semiconductors. The doping of the former involves the oxidation or reduction of the conjugated backbone, while that of the latter arises from the ionization of the doping elements. Doping does not produce any new band in inorganic semiconductors. But new energy levels or bands are generated in conducting polymers after doping because the charge carriers including electrons and holes in doped inorganic semiconductors can be considered as free electrons while the charge carriers of conducting polymers strongly couple with the lattice vibration.

Because of the strong coupling between the electrons and lattice vibration, the charge carriers in conducting polymers are related to their symmetry. The charge carriers in doped polyacetylene are solitons because of its high symmetry in the chemical structure. A soliton has a positive charge when it is oxidative doping of polyacetylene, and the charge is negative when it is reductive doping (Figure 1.9). The solitons are delocalized along the conjugated polymer backbone. As shown in Figure 1.10, there is a discrete soliton level between the conduction band and valence band at low doping level. There is no electron on the soliton level if oxidative doping occurs for polyacetylene, and there are two electrons on the soliton level when it is reductive doping. When the doping degree is high enough, the solitions couple with each other and the soliton level becomes a continuous soliton band.

The symmetry of the chemical structure of other conducting polymers such as polypyrrole (PPy), polythiophene (PTh), and polyaniline (PANi) is lower than that of polyacetylene. Their charge carriers are polarons at low doping degree. As shown in Figure 1.11, there is a positive charge and an electron for a polaron. Because a polaron mainly involves two atoms, there are two polaron levels between the conduction band and valence band (Figure 1.12). The electron is on the lower polaron level. At the medium

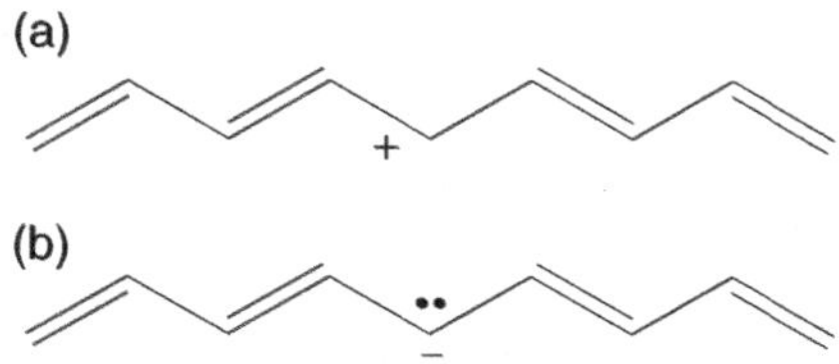

Figure 1.9 Polyacetylene (PA) with (a) a positive soliton and (b) a negative soliton.

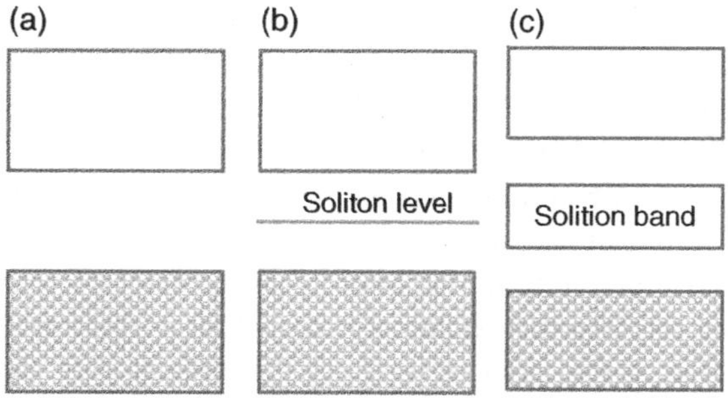

Figure 1.10 Structures of polyactylene in (a) neutral state, (b) low doping degree and, (c) high doping degree.

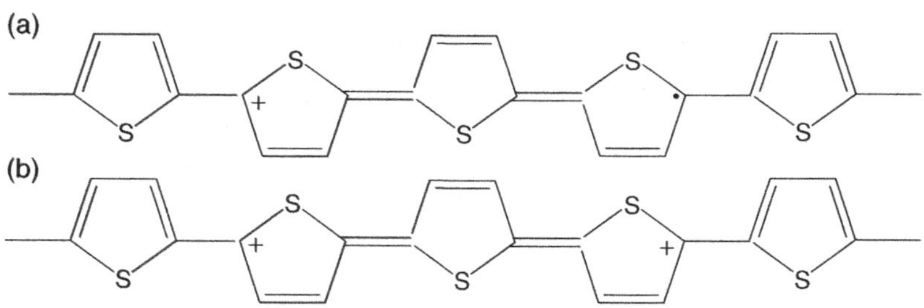

Figure 1.11 Polythiophene with (a) a polaron and (b) a bipolaron.

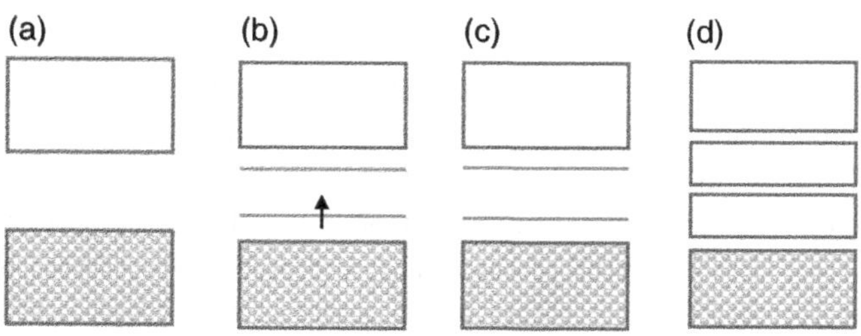

Figure 1.12 Band structures of conducting polymers with less symmetry in (a) neutral state, (b) low doping degree, (c) medium doping degree, and (d) high doping degree. There is one electron on the lower polaron level in (b), while no electron on the bipolaron levels and bipolaron bands in (c) and (d).

doping degree, the polarons can interact with each other, and two polarons become a bipolaron. A bipolaron has two positive charges and no electron. At high doping degree, the bipolarons couple each other, and the two bipolaron levels turn to two continuous bipolaron bands.

1.1.5.2 Charge Carrier Mobility

The conduction electrons in metals are the valence electrons of metal atoms. The conduction electrons can be scattered by the lattice consisted of metal cations, and the lattice scattering lowers the electron mobility. Because of the much higher ion concentration in metals than in inorganic semiconductors, metals usually have a much lower charge carrier mobility than inorganic semiconductors. The charge carrier mobility of inorganic semiconductors is mainly affected by impurities, defects, and lattice vibration. A doping atom can be considered as an impurity, and it forms a scattering center for the charge carriers. As a result, doping lowers the charge carrier mobility. Therefore, the charge carrier mobility of inorganic semiconductor and conducting polymers usually decreases with the increasing doping level.

Different from metals and inorganic semiconductor that can be prefect crystals, a conducting polymer cannot have 100% crystallinity. There are both amorphous and crystalline domains in conducting polymers (Figure 1.13). The amorphous domain has a conductivity much lower than that of the crystalline domain. The charge transports include intrachain, interchain, and inter-domain processes. The interchain charge transport is usually the dominant process. Thus, the charge carrier

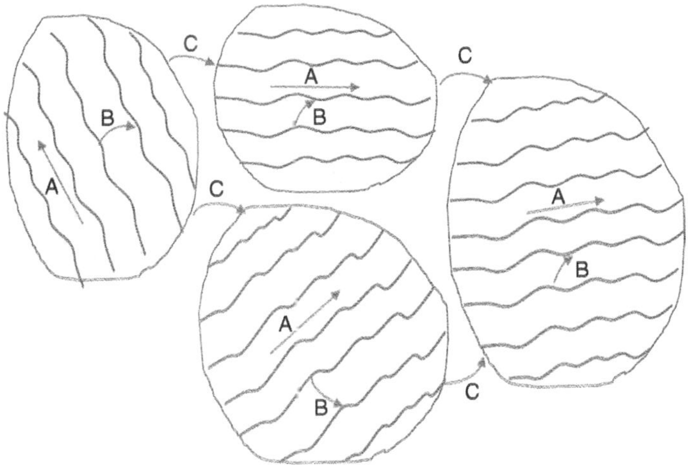

Figure 1.13 (a) Intrachain, (b) interchain, and (c) inter-domain charge transports in conducting polymers.

mobility of conducting polymers strongly depends on the crystallinity. High crystallinity can facilitate the interchain charge transport, and thus give rise to high charge carrier mobility. A conducting polymer with the same chemical composition can have conductivities spanning several orders by magnitude.

The crystallinity of some conducting polymers can be increased through post treatment. As a result, the conductivity can be enhanced by several orders of magnitude. When a chemical is used for the post treatment, the post treatment method is usually called "secondary doping." Secondary doping has been extensively reported for solution-processable conducting polymers including polyaniline and PEDOT:PSS [15–19]. Polyaniline doped with a large counter anion like camphorsulfonic acid (HCSA) can be dispersed in organic solvent such as *m*-cresol and chloroform [20]. The solvent significantly affects the crystallinity and the conductivity of polyaniline:HCSA (PANi:HCSA) because it can affect the conformation of the polymer chains in solution. When a PANi:HCSA film is prepared from chloroform, its conductivity is only 10^{-1} S cm^{-1}, while it can be >100 S cm^{-1} when it is obtained from *m*-cresol. This difference in conductivity is attributed to the effect of the solvents on the crystallinity of polyaniline. As shown in Figure 1.14, there is a strong absorption in the infrared range for the PANi:HCSA film cast from its *m*-cresol solution [15]. This suggests a high crystallinity and the delocalization of the charge carrier. Instead, only a band appears in the near infrared (NIR) range when chloroform is used as the solvent. This is consistent with the low crystallinity.

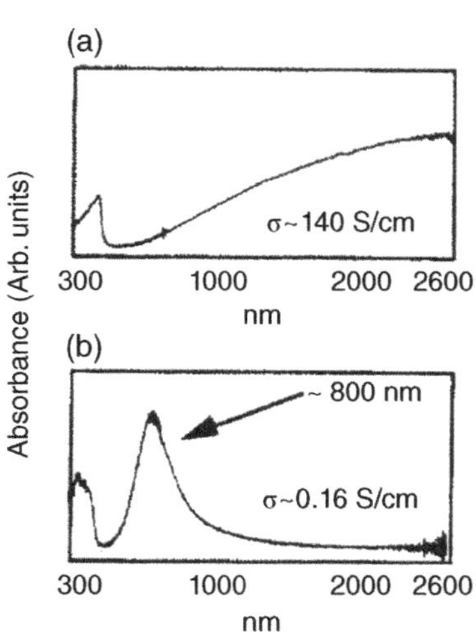

Figure 1.14 UV–visible–NIR absorption spectra of PANi:HCSA films casted from solutions with (a) *m*-cresol or (b) chloroform as the solvent, respectively. *Source:* MacDiarmid and Epstein [15]. © Elsevier.

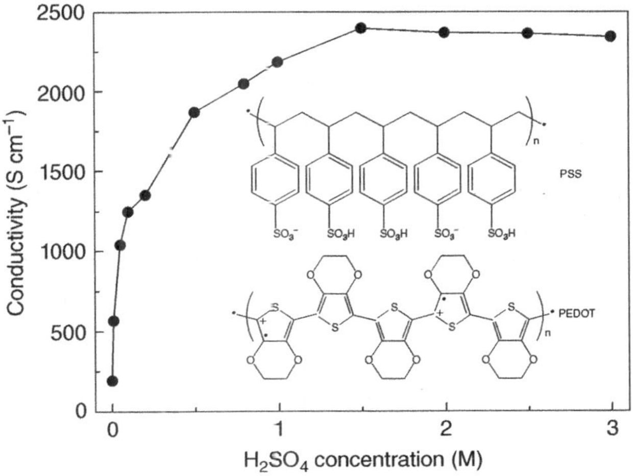

Figure 1.15 Conductivity of PEDOT:PSS after secondary doping with H_2SO_4. *Source:* Xia et al. [21]. © John Wiley & Sons.

A couple of secondary doping methods were also reported for PEDOT:PSS. PEDOT:PSS can be dispersed in water [17–19]. The conductivity of an as-prepared PEDOT:PSS film is only 10^{-1} S cm^{-1}, and it can be significantly enhanced by secondary doping. The chemicals for the secondary doping can be organic solvents like dimethyl sulfoxide (DMSO), ethylene glycol (EG) and even methanol, cosolvent of water and organic solvents, aqueous or organic solutions, ionic liquids, and acids. Particularly, the conductivity of PEDOT:PSS can be enhanced from 10^{-1} S cm^{-1} to higher than 3000 S cm^{-1} through a post acid treatment [21–24]. The conductivity enhancement is ascribed to the partial removal of the insulating PSSH from PEDOT:PSS and the conformational change of PEDOT from coil to expanded-coil or linear structure. Less PSSH chains and linear PEDOT conformation can lead to high crystallinity and thus high conductivity.

Mechanical stretching can orient polymer chains along the tensile direction and thus increase the crystallinity. It has been used to increase the conductivity of conducting polymers as well [25, 26]. For example, it can increase the conductivity of polyacetylene up to 10^5 S cm^{-1}.

1.1.5.3 Temperature Dependence of Conductivity

Temperature dependence of the conductivity is often used to study the charge transport mechanism because metals and semiconductors have different temperature dependences of their conductivity. Conducting polymers are considered as disordered systems. Their charge transport is dominated by interchain charge hopping, that is different from those of both metals and semiconductors.

1.1.5.3.1 Metals The temperature dependence of the conductivity is related to the charge transport mechanism. Varying the temperature does not affect the charge carrier density but the charge carrier mobility of metals. As the lattice vibrations become stronger at higher temperature, the scattering of the conduction electrons by lattice vibrations becomes more significant at higher temperature. This lowers the charge carrier mobility. The conductivity of metals thus decreases with the increasing temperature. There are two empirical formulas for the temperature dependence of the resistance of metals. One is the power law form,

$$\rho = \rho_0 \left(\frac{T}{T_0} \right)^n . \tag{1.15}$$

The resistivity (ρ) at temperature (T) is related to the resistivity (ρ_0) at the reference temperature (T_0). The resistivity linearly increases with temperature in another form, and the temperature effect is characterized by the temperature coefficient of resistivity (α_0),

$$\rho = \rho_0 \left[1 + \alpha_0 \left(T - T_0 \right) \right] . \tag{1.16}$$

In both forms, the resistivity of metals increases with the increasing temperature.

1.1.5.3.2 Semiconductors The temperature dependence of the conductivity of semiconductors is quite different from metals. The temperature can affect both the charge carrier density and the charge carrier mobility of semiconductors. There are three temperature ranges in terms of the temperature effect on the charge carrier density. At the low temperature range below the saturation temperature (T_s), the thermal energy is less than the ionization energy of the doping atoms. As a result, not all the doping atoms ionize. The electron (or hole) density in the conduction band (or valence band) depends on the thermal excitation of the electrons (or holes) from the dopant to the conduction band (or valence band). The charge carrier density increases with the increasing temperature. In the medium temperature of $T_s < T < T_i$ with T_i being the intrinsic temperature, the thermal energy is high enough for the ionization of all the dopants, but it is lower than the energy bandgap between the valence band and the conduction band. The charge carrier density equals the dopant density, and it is constant in this temperature range. At the high temperature range of $T > T_i$, the thermal energy can excite the electrons from the conduction band to the valence band. This produces electrons in the conduction band and holes in the valence band. In this high temperature range, both the electron and hole densities increase with the increasing temperature.

There are two temperature regimes for the temperature dependence of the charge carrier mobility of semiconductors. At low temperature regime, the ionized impurities are the charge carrier scattering centers. In this temperature range, the charge carrier mobility increases with the increasing temperature because the scattering area decreases. At high temperature, the scattering by the lattice vibration becomes the dominant factor for the charge carrier mobility. Similar to metals, the charge carrier mobility decreases with the increasing temperature in the high temperature regime.

Therefore, the temperature dependence of the conductivity of semiconductor is more complicated than that of metals. At low and high temperature range, the conductivity of semiconductor increases with the increasing temperature. The conductivity can be less sensitive to the temperature or even decrease in the medium temperature range.

The temperature dependence of the conductivity is usually used to determine whether a conductor is metallic or semiconductive. In general, if the conductivity decreases with the increasing temperature, it is metallic. In contrast, if the conductivity increases with the increasing temperature, it is considered as semiconductive behavior. This is also frequently used to judge the charge transport mechanism of non-conventional materials, such as carbon nanotubes, graphene, and charge-transfer organic salts.

1.1.5.3.3 Conducting Polymers The temperature dependence of the conductivity of conducting polymers is different because they are disordered systems. The charge transport is dominated by the interchain charge hopping. In general, the conductivity of conducting polymers follows the one-dimensional variable range hopping model,

$$\rho = \rho_0 \exp\left[\left(\frac{T_0}{T}\right)^{1/2}\right]. \tag{1.17}$$

In terms of this model, the resistivity increases with the increasing temperature. The kT_0 value suggests the energy barrier for the charge hopping. It decreases with the increasing crystallinity. It is easy to confuse the charge hopping mechanism with the metallic band structure of conducting polymers. At high doping level, the conducting polymers can have energy band

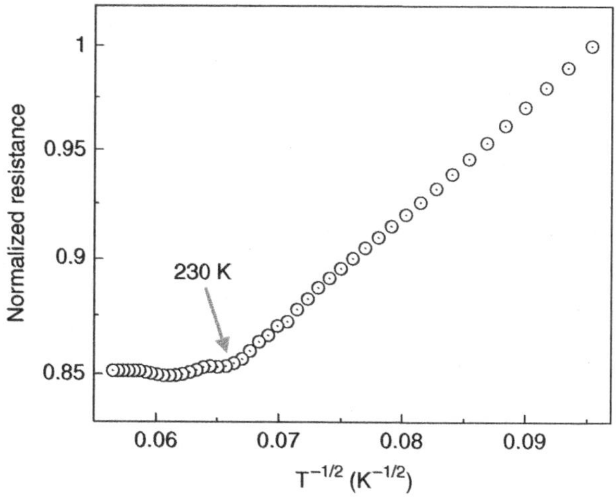

Figure 1.16 Temperature dependence of the resistance of a PEDOT:PSS treated with H_2SO_4. *Source:* Xia et al. [21]. © John Wiley & Sons.

structure like metals. This picture is applicable for an individual chain. For a conducting polymer sample, the charge transport is dominated by interchain charge hopping.

However, semimetallic or metallic behavior, that is, the conductivity is insensitive to temperature or decreases with the increasing temperature, was reported on a few conducting polymers with high conductivity. For example, H_2SO_4 treatment can enhance the conductivity of a PEDOT:PSS film prepared from the Clevious PH1000 aqueous dispersion from ~0.2 $S\,cm^{-1}$ to >3000 $S\,cm^{-1}$ [21]. As shown in Figure 1.16, when the temperature is lower than 230 K, the temperature dependence of the resistance still follows the one-dimensional variable range hopping model. Nevertheless, when the temperature is higher than 230 K, the thermal energy can overcome the energy barrier for the interchain charge hopping. As a result, the resistance becomes insensitive to temperature. This indicates semimetallic or metallic behavior. Metallic or semimetallic behavior was also observed on highly conductive polyaniline [27].

1.1.5.4 Conductivity of Composites

Some composites particularly the dispersion of inorganic nano-fillers in polymer matrix can exhibit high thermoelectric properties [28, 29]. There can be three different structures for a two-phase composite (Figure 1.17). The resistivity (ρ_c) of the composites depends on the volume fractions and the resistivity of the two phases. When the two phases α and β are in series, the resistivity of the composite is given by,

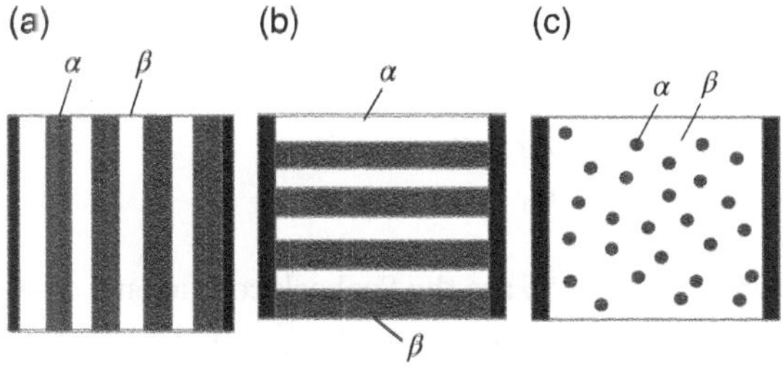

Figure 1.17 Structures of a composite with two phases of α and β (a) in series, (b) in parallel, and (c) one phase dispersed in another phase.

$$\rho_c = \chi_\alpha \rho_\alpha + \chi_\beta \rho_\beta, \tag{1.18}$$

where χ_α and χ_β are the volume fractions of the α and β phases, respectively, and ρ_α and ρ_β are the resisitivities of the two phases, respectively. The resistivity of the composite is dominated by the phase of higher resistivity.

When the two phases of α and β are parallel, the conductivity (σ_c) of the composites is related to the conductivities of the two phases by the following equation,

$$\sigma_c = \chi_\alpha \sigma_\alpha + \chi_\beta \sigma_\beta. \tag{1.19}$$

The phase with higher conductivity will be the dominant one for the conductivity of the composite.

If a composite has a structure of the α phase dispersed in the β phase, the α phase is the dispersed phase and the β phase is the matrix. The resistivity of the composite depends on the relative resistivities of the two phases. If the α phase is more resistive than the β phase, $\rho_\alpha > 10\, \rho_\beta$, the resistivity of the composite is given by

$$\rho_c = \rho_\beta \frac{1 + \frac{1}{2}\chi_\alpha}{1 - \chi_\alpha}, \tag{1.20}$$

If the α phase is less resistive than the β phase, $\rho_\alpha < (1/10)\rho_\beta$, the resistivity of the composite is given by

$$\rho_c = \rho_\beta \frac{1 - \chi_\alpha}{1 + 2\chi_\alpha}. \tag{1.21}$$

The Seebeck coefficient of the composites is also related to the microstructure of the two phases. When the α and β phases are in series, the Seebeck coefficient (S_c) of the composite is given by,

$$S_c = \frac{\frac{\chi_\beta S_\beta}{\kappa_\beta} + \frac{\chi_\alpha S_\alpha}{\kappa_\alpha}}{\frac{\chi_\beta}{\kappa_\beta} + \frac{\chi_\alpha}{\kappa_\alpha}}, \tag{1.22}$$

where $S\alpha$ and $S\beta$ are the Seebeck coefficients of the α and β phases, respectively, and $\kappa\alpha$ and $\kappa\beta$ are the thermal conductivities of the two phases, respectively. When the two phases are in parallel, the Seebeck coefficient of the composite is given by

$$S_c = \frac{\chi_\beta \sigma_\beta S_\beta + \chi_\alpha \sigma_\alpha S_\alpha}{\chi_\beta \sigma_\beta + \chi_\alpha \sigma_\alpha}. \tag{1.23}$$

1.1.6 Thermal Conductivity

The thermal conductivity (κ) of an electronic material include the lattice thermal conductivity (κ_L) and the electronic thermal conductivity (κ_e),

$$\kappa = \kappa_L + \kappa_e. \tag{1.24}$$

The lattice thermal conductivity is given by the Debye equation,

$$\kappa_L = \frac{1}{3} c_p v l, \tag{1.25}$$

where c_p, v, and l are the specific heat capacity, the phonon velocity, and the phonon mean free path, respectively. The phonon mean free path of polymers is usually extremely small due to the scattering by other phonons, defects, and grain boundaries. Thus, polymers usually have a thermal conductivity much lower than metals and semiconductors. For example, the values of c_p, v, and l are 1.8×10^6 J (K^{-1} m^{-3}), 1.58×10^3 m s^{-1}, and 0. nm, respectively, for PEDOT:PSS. The κ_L value of PEDOT:PSS is thus ~0.3 W (m^{-1} K^{-1}). The lattice thermal conductivity is related to the crystallinity of polymers. The higher the crystallinity, the higher the thermal conductivity.

The electronic thermal conductivity is related to the electrical conductivity by the Wiedemann–Franz law,

$$\kappa_e = L\sigma T, \tag{1.26}$$

where L is the Lorentz number. The Sommerfeld value of the Lorenz number at $T \rightarrow 0$ K is given by

$$L_0 = \frac{\pi^2 k_3^2}{3e^2}. \tag{1.27}$$

One important reason for the thermoelectric application of polymer nanocomposites is their low thermal conductivity. The thermal conductivity of a composite with polymer matrix and fillers depends on its microstructure [30]. A composite has the highest thermal conductivity when its structure is modeled with two components in parallel and the lowest thermal conductivity when it is modeled with the two components in series.

The thermal conductivity (κ_c) of a composite by the parallel model is given by

$$\kappa_c = \chi_\alpha \kappa_\alpha + \chi_\beta \kappa_\beta, \tag{1.28}$$

where χ_α and χ_β are the volume fractions of the two phases, respectively, and κ_α and κ_β are the thermal conductivities of the two phases, respectively. The parallel model usually overestimates the contribution of the fillers to the overall thermal conductivity of the composites.

The thermal conductivity of a composite by the series model is given by

$$\frac{1}{\kappa_c} = \frac{\chi_\alpha}{\kappa_\alpha} + \frac{\chi_\beta}{\kappa_\beta}. \tag{1.29}$$

The thermal conductivity of composites with nano-fillers dispersed in polymers is between that by the parallel model and series model, and it is usually close to the value by the series model.

A more precise equation for the thermal conductivity of composites with fillers in matrix is given by the Maxwell–Garnett equation,

$$\kappa_c = \kappa_m \left[1 + \frac{3\chi_f (\delta - 1)}{2 + \delta - \chi_f (\delta - 1)} \right], \tag{1.30}$$

where κ_m and κ_f are the thermal conductivities of the matrix and fillers, respectively, and $\delta = \kappa_f/\kappa_m$. The thermal conductivity of nanofillers like carbon nanotubes and graphene is usually much larger than that of the polymer matrix. Assuming that the values for a composite are $\delta = 1$ and $\chi_f = 0.05$, we can obtain $\kappa_c = 1.1\ \kappa_m$ in terms of the Eq. (1.32), that is, the thermal conductivity of the composite is almost the same as the polymer matrix. Thus, the composite may have high electrical conductivity and high Seebeck coefficient but thermal conductivity is low. In terms of the Maxwell-Garnett equation, the thermal conductivity of a composite linearly increases with the filler loading when the filler loading is low.

When the fillers form percolation for thermal conduction, the thermal conductivity of composites is underestimated by the Maxwell–Garnet equation. At the filler concentration higher than the percolation threshold, the thermal conductivity is given by the Bruggeman equation,

$$\chi_m \frac{\kappa_m - \kappa_c}{\kappa_m + 2\kappa_c} + \chi_f \frac{\kappa_f - \kappa_c}{\kappa_f - 2\kappa_c} = 0. \qquad (1.31)$$

There is an interface between the polymer matrix and fillers in composites. Phonons can be scattered at this interface if the interfacial coupling between the matrix and fillers is weak. If the fillers are small and the separation between the fillers is large, the thermal conductivity of the composites is low.

1.2 Thermoelectric Generators

1.2.1 Dependence of Thermoelectric Efficiency on ZT

The most important application for thermoelectric materials is the energy conversion between heat and electricity. It is important to understand the heat transfer, drift electrical current and thermoelectric current in thermoelectric generators. The Seebeck effect is exploited to convert heat into electricity in thermoelectric generator. Assume that the p-type and n-type thermoelectric materials have the same ZT value, the thermoelectric conversion efficiency depends on the ZT value of the thermoelectric materials,

$$\eta = \eta_c \frac{\sqrt{1 + Z\overline{T}} - 1}{\sqrt{1 + Z\overline{T}} + \frac{T_C}{T_H}}, \qquad (1.32)$$

where η_c is the Carnot efficiency for the temperature difference at the hot (T_H) and cold ends (T_C),

$$\eta_c = \frac{T_H - T_C}{T_H}, \qquad (1.33)$$

and $\overline{T}$ is the mean value of the operation temperatures,

$$\overline{T} = \frac{T_H + T_C}{2}. \qquad (1.34)$$

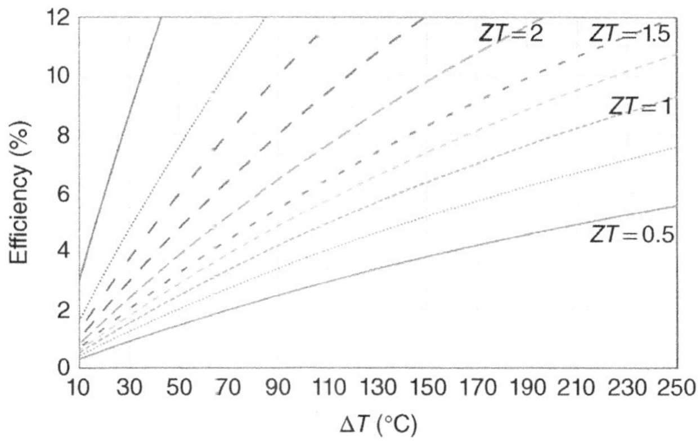

Figure 1.18 Dependence of thermoelectric generation efficiency on the temperature and *ZT* value. The top thick line is the Carnot efficiency. *Source:* Fanciulli et al. [31]. © Elsevier.

A higher *ZT* value gives rise to a higher conversion efficiency (Figure 1.18). But the thermoelectric conversion efficiency is always lower than the Carnot efficiency.

1.2.2 Effect of Electrical and Thermal Contact Resistances On Thermoelectric Performance

Apart from the *ZT* value of the thermoelectric materials, other factors can affect the thermoelectric efficiency as well. In addition to the thermoelectric materials, electrodes, heat sink, and cold sink are needed in thermoelectric generators. In a thermoelectric module, a *p*-type thermoelectric leg is connected to an *n*-type thermoelectric leg through a metallic contact layer, that is, electrode (Figure 1.19). Thus, there are interfaces between the thermoelectric materials and the contact layers. They introduce both electrical and thermal contact resistances. This lowers the thermoelectric performance of the module because of the Joule heat generation and the temperature drop at the contacts. For a thermoelectric leg with two contact layers at the two ends, the effective resistance (R_e) will be increased by the contact resistance (R_c),

$$R_e = R + 2R_c, \tag{1.35}$$

where R is the resistance of the thermoelectric leg. The temperature drop (ΔT_c) at the contacts lowers the actual temperature difference (ΔT) on the thermoelectric leg,

$$\Delta T = \Delta T_0 - 2\Delta T_c, \tag{1.36}$$

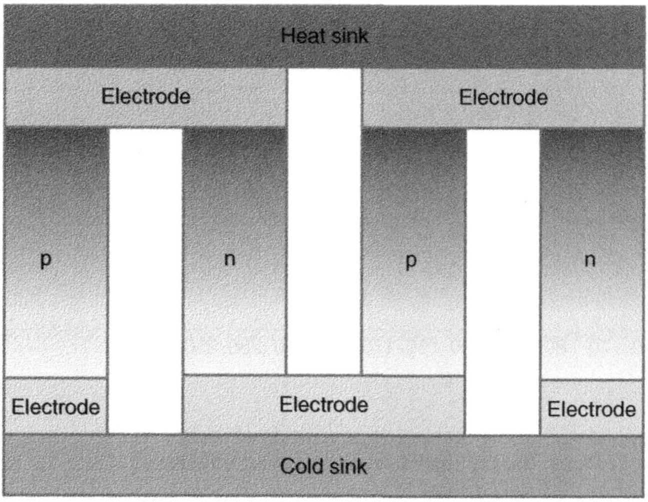

Figure 1.19 Schematic structure of a thermoelectric module with *n*-type and *p*-type thermoelectric legs.

where ΔT_0 is the temperature difference between the heat sink and the cold sink.

The deterioration of the thermoelectric performance by the electrical and thermal contact resistances can be modeled into their effects on the effective ZT value. If a thermoelectric leg with a length of l and figure of merit of ZT is connected to the contact layers with the conductivity of (σ_{con}) and thermal conductivity of (κ_{con}), the effective ZT_e value will decrease,

$$ZT_e = \frac{ZT}{\left(1+\frac{2\sigma}{\sigma_{con} l}\right)\left(1+\frac{2\kappa}{\kappa_{con} l}\right)}, \tag{1.37}$$

where σ and κ are the electrical conductivity and thermal conductivity of the thermoelectric material, respectively. Thus, the decrease in the ZT value is related to the electrical conductivities and thermal conductivities of the contact layer and the thermal electric leg and the length of the thermoelectric leg. An effective material for the contact layer should have high electrical conductivity and high thermal conductivity. It is important to make the thermoelectric interfaces with small electrical and thermal contact resistances. To design high thermoelectric conversion efficiency, the contact area between the electrodes and thermoelectric legs and the length of the thermoelectric legs are important parameters [32, 33].

In addition, the thermal expansion coefficient of the contact layers should match with the thermoelectric legs particularly for high temperature application. Otherwise, the thermal expansion mismatching can produce fractures or cracks at the interfaces and thus increase the electrical and thermal contact resistances.

The energy conversion efficiency of a thermoelectric generator or module is the electrical power output divided by the heat power input. But it is difficult to accurately determine the heat power input. Thus, electrical powers instead of the efficiency values are usually provided in literature. The electrical power output depends on the temperature difference (ΔT). Because of the temperature dependences of the Seebeck coefficient, the ZT value and the resistances of thermoelectric legs and electrodes, the open-circuit voltage and the total electrical resistance vary with ΔT as well. In addition, cracks and materials degradation can happen under thermal cycling. Thus, the evolutions of the power output, open-circuit voltage and total electrical resistance are sometimes presented with the number of the thermal cycles.

1.2.3 Equation of Thermoelectric Efficiency

The equation for the thermoelectric efficiency of thermoelectric generators is derived as below. The electrical current density through a thermoelectric material includes the drift current density (J_d) driven by the potential gradient and the thermal diffusion current density (J_t) driven by the temperature gradient,

$$J = J_d + J_t = -\sigma \frac{\partial V}{\partial x} - \sigma S \frac{\partial T}{\partial x}, \tag{1.38}$$

where V is the potential, T is the temperature, and σ and S are the electrical conductivity and the Seebeck coefficient of the thermoelectric material, respectively.

When a thermoelectric leg is disconnected with the external circuit, the voltage on the thermoelectric leg is the open-circuit voltage (V_{OC}) (Figure 1.20a). At this open-circuit condition, the net current at any position is zero, $J = 0$. The drift current density is opposite to the thermal diffusion current density, and they cancel each other at any position. In terms of the Eq. (1.38), the open-circuit voltage is given by

$$\Delta V_{OC} = -S(T_H - T_C). \tag{1.39}$$

When an external load is connected to the thermoelectric leg, the charge carriers can flow through the external circuit. The current density increases

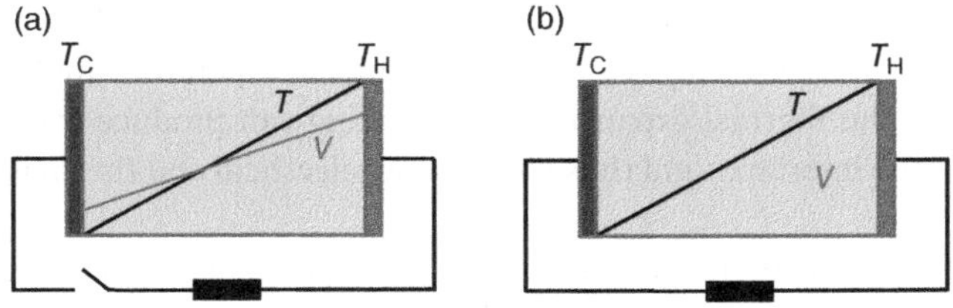

Figure 1.20 Steady temperature and voltage profiles of a thermoelectric leg under (a) open-circuit and (b) short-circuit conditions.

with the decreasing external load resistance. When the external load resistance is zero, the voltage is zero. This is the short-circuit condition. The current density equals the thermal diffusion current density J_t (Figure 1.20b).

It should be noted that ΔT is lowered at the short circuit condition in comparison with the open-circuit condition because of the Peltier effect. According to the Peltier effect, the current flow cools down the hot side and heats up the cold side. This will lower the J_t.

In the normal case, there are both drift current density and thermal diffusion current density. In terms of the conservation law, the total amount of charge carriers in a material must be conserved. Thus, the variation of the charge carrier density (n) over time is related to the gradient of the current density,

$$q\frac{dn}{dt} = -\frac{\partial J}{\partial x}. \tag{1.40}$$

By replacing J with the Eq. (1.38), it is given,

$$q\frac{dn}{dt} = -\sigma\frac{\partial^2 V}{\partial x^2} - \sigma S\frac{\partial^2 T}{\partial x^2}. \tag{1.41}$$

This indicates that the variation of charge carrier density over time can be achieved by varying the voltage or the temperature gradient over position.

The energy flux in a thermoelectric material includes the heat transport and thermoelectric conversion. J_Q is used for the heat flux that is the amount of energy crossing a cross-section area per unit time,

$$J_Q = -\kappa\frac{\partial T}{\partial x} + STJ. \tag{1.42}$$

The first term in the right side of Eq. (1.42) is the heat conduction from the hot to cold part within a material. The second term is the heat related to the thermoelectric effects. Derivation of the Eq. (1.42) with position leads to

$$\frac{\partial J_Q}{\partial x} = -\frac{\partial}{\partial x}\left(\kappa\frac{\partial T}{\partial x}\right) + TJ\frac{\partial S}{\partial x} + SJ\frac{\partial T}{\partial x} + ST\frac{\partial J}{\partial x}. \tag{1.43}$$

It is important to understand the terms in the right side of the Eq. (1.43). The second term where S varies with position is related to the Peltier and Thomson effects. As shown in Figure 1.21a, the system has a uniform temperature, and there is current flow through the system. This can be the case, right after applying a current to a system which was initially in equilibrium. Because the Seebeck coefficient (S_M) of the metal electrode is different from that (S) of the thermoelectric material, the transport of the current from the left metal electrode through the contact into the thermoelectric material leads to the energy absorption (cooling) by (S_M-S)TJ due to the Peltier effect. On the other hand, the same amount of energy is released at the right contact, when the current flows from the thermoelectric material into the right metal electrode. The energy release is also owing to the Peltier effect.

Figure 1.21b shows the case when the temperature is not uniform. In addition to the Peltier effect at the two contacts, STJ is not uniform inside the thermoelectric material because S varies with temperature. The Seebeck coefficient (S_a) at the position a is different from that (S_b) at the position b. As a result, heat transfer takes place between the two positions because of the Thomson effect. The heat transfer (Q_T) due to the Thomson effect is given by

$$Q_T = -T\frac{dS}{dT}J\frac{\partial T}{\partial x}. \tag{1.44}$$

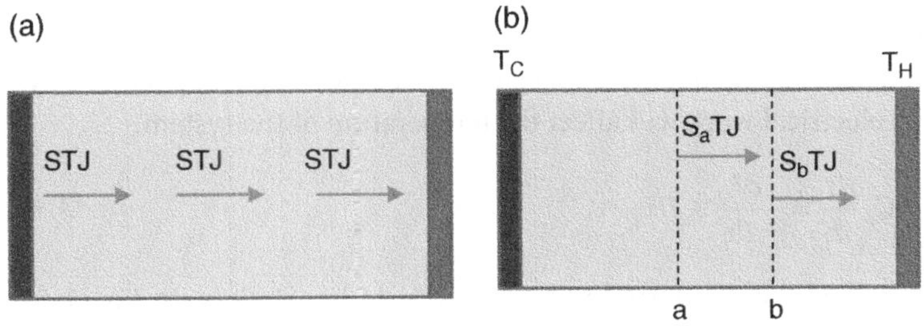

Figure 1.21 Scheme of the different heat fluxes of an n-type thermoelectric material connected to metallic contacts under the flow of an electrical current density J. (a) Only Peltier effect takes place at the junctions and there is no temperature difference. (b) Under a temperature gradient, S varies inside the thermoelectric material, which produces heat generation in the bulk (Thomson effect), the thermoelectric effects at the positions a and b are indicated. Peltier effect also takes place at the junctions.

The third term $SJ\partial T/\partial x$ in the right side of Eq. (1.43) can be converted to the potential gradient in terms of the Eq. (1.38),

$$SJ\frac{\partial T}{\partial x} = -J\frac{\partial V}{\partial x} - \frac{J^2}{\sigma} = JE - \frac{J^2}{\sigma}, \tag{1.45}$$

where $E = -\frac{\partial V}{\partial x}$ is the electric field. The first term in the right side of Eq. (1.45) is the electrical work, and the second term is the Joule heat generated in the material by the electrical current.

The last term $ST\frac{\partial J}{\partial x}$ at the right side of Eq. (1.43) is the energy variation due to the variation of the charge carrier concentration. At steady-state condition the current becomes uniform, and this term is thus zero.

These analyses indicate that the energy flux variation in a material can include the variations of heat conduction, the Peltier effect, the Thomson effect, the Joule heating, the electrical work, and the distribution of charge carriers,

$$\frac{\partial J_Q}{\partial x} = -\frac{\partial}{\partial x}\left(\kappa\frac{\partial T}{\partial x}\right) + T\frac{dS}{dT}J\frac{\partial S}{\partial x} - \frac{J^2}{\sigma} + JE + ST\frac{\partial J}{\partial x}. \tag{1.46}$$

Heat flux can change the temperature of a material. Consider in the volume of Adx with A as the cross-section area,

$$dC_p\frac{dT}{dt} = -\frac{\partial J_Q}{\partial x}, \tag{1.47}$$

where d is the mass density and C_p is the specific heat.

When a material is initially at thermal equilibrium with homogeneous temperature, its temperature can be varied only when an electrical work or heat transfer is applied to it. If an external electric field is applied to a system, the electrical work will affect the temperature of the system,

$$dC_p\frac{dT}{dt} = -\frac{\partial J_Q}{\partial x} + JE. \tag{1.48}$$

By replacing the first term in the right side of Eq. (1.48) with Eq. (1.46), the following equation is obtained,

$$dC_p\frac{dT}{dt} = \frac{\partial}{\partial x}\left(\kappa\frac{\partial T}{\partial x}\right) - T\frac{dS}{dT}J\frac{\partial S}{\partial x} + \frac{J^2}{\sigma} - ST\frac{\partial J}{\partial x}. \tag{1.49}$$

Assume that the thermal conductivity is independent of position at steady state, (1.49) becomes

$$\kappa\frac{\partial^2 T}{\partial x^2} - T\frac{dS}{dT}J\frac{\partial S}{\partial x} + \frac{J^2}{\sigma} = 0, \tag{1.50}$$

as $\frac{dT}{dt} = 0$ and $\frac{\partial J}{\partial x} = 0$ at steady state. The Eq. (1.50) is used for the evaluation of thermoelectric efficiency.

The input heat flux or the temperature difference between the hot and cold sides is usually constant when a thermoelectric system is evaluated. Consider that the material properties are independent of the temperature and they are identical for the *p*-type and *n*-type legs, the power delivered to the external load is given by

$$P = I\Delta V_{ex}, \tag{1.51}$$

where I is the current through the external load and $\Delta V_{ex} = IR_{ex}$ is the voltage drop on the external load.

The values of I, ΔV_{ex}, and P depend on the resistance of the external load. The resistance of the external load is usually varied from the open-circuit voltage condition ($R_{ex} \to \infty$) to the short-circuit voltage ($R_{ex} \to 0$) to find the optimal power on the external load.

As shown in Figure 1.22, for a system with a *p*-type leg and an *n*-type leg, the open-circuit voltage is

$$V_{OC} = 2|S|\Delta T, \tag{1.52}$$

when the *p*- and *n*-type thermoelectric materials have the same absolute Seebeck coefficient value. When the system is connected to an external load

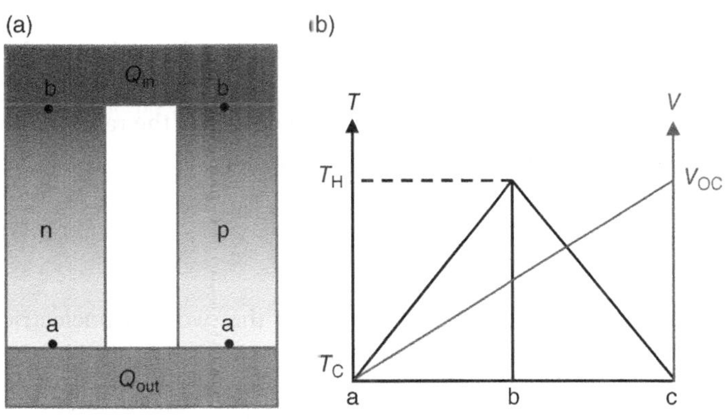

Figure 1.22 (a) A thermoelectric generator with *n*- and *p*-type legs. Heat transfer (Q_{in}) into the thermoelectric legs from the hot side and heat transfer (Q_{out}) out from the cold side. (b) Temperature and voltage profiles at open-circuit condition.

and form a close circuit, the open-circuit voltage dissipated on the external load and the internal resistance (R_{in}) that includes the resistances of materials, contacts, and wires,

$$2|S|\Delta T = IR_{ex} + IR_{in} \tag{1.53}$$

The current through the circuit can be obtained as

$$I = \frac{2|S|\Delta T}{R_{in} + R_{ex}}. \tag{1.54}$$

Thus, the equation for the power is given by

$$P = I\Delta V_{ex} = \frac{\left(2|S|\Delta T\right)^2}{\left(R_{in} + R_{ex}\right)^2} R_{ex}. \tag{1.55}$$

By introducing a parameter $m = \frac{R_{ex}}{R_{in}}$, the equation (1.55) becomes

$$P = \frac{\left(2|S|\Delta T\right)^2}{R_{in}} \frac{m}{\left(1+m\right)^2}. \tag{1.56}$$

When $R_{in} = R_{ex}$, the external load has the maximum power (P_{max}),

$$P_{max} = \frac{\left(2|S|\Delta T\right)^2}{4R_{in}}. \tag{1.57}$$

Figure 1.23 shows the variations of the output power with the external load resistance at different temperature differences [34]. The output power increases when the temperature difference increases. However, at the optimal output power, the corresponding load resistance is independent of the temperature difference.

The equation for the efficiency (η) can be obtained as it is the ratio of the output power to the incident heat (Q_{in}) into the system,

$$\eta = \frac{P}{Q_{in}}. \tag{1.58}$$

Heat energy transport through this system with the two thermoelectric legs, p- and n-type legs. The incident heat at the hot side is balanced by the Peltier effect at the contacts between the legs and the metal electrodes, and the heat transfers due to the temperature gradient. The heat transferring from the hot to the cold side is equivalent to the Peltier effect at the contacts

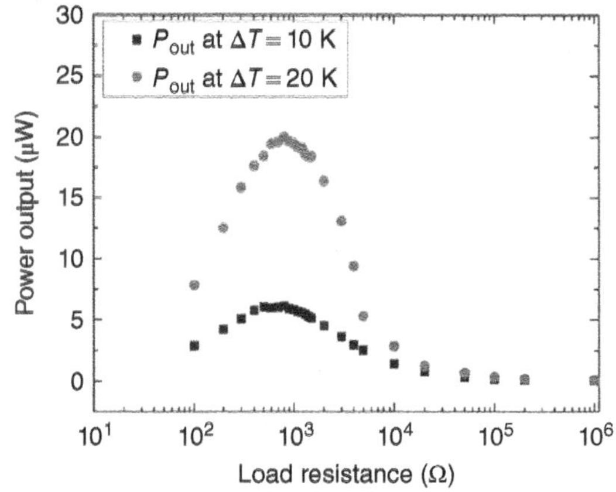

Figure 1.23 Variations of the power output of a TEG with the load resistance at $\Delta T = 10$ and 20 K. *Source:* Madan et al. [34]. © American Chemical Society.

and the heat out from the cold side. Thus, the equations at the two boundaries (hot and cold side) are given by

$$\frac{Q_{in}}{A} - 2\frac{ST_H I}{A} + 2\kappa\left(\frac{\partial T}{\partial x}\right)_0 = 0, \text{ at } x = 0, \text{ hot side,} \tag{1.59}$$

$$-\frac{Q_{out}}{A} + 2\frac{ST_C I}{A} - 2\kappa\left(\frac{\partial T}{\partial x}\right)_L = 0, \text{ at } x = L, \text{ cold side.} \tag{1.60}$$

By ignoring the Thomson effect, the equation (1.50) for one leg becomes

$$\kappa\frac{\partial^2 T}{\partial x^2} + \frac{I^2}{A^2\sigma} = 0. \tag{1.61}$$

Using the boundary conditions of $T(0) = T_H$ and $T(L) = T_C$, the following equations can be obtained,

$$T(x) = -\frac{I^2 x^2}{2\kappa A^2\sigma} + \left(\frac{I^2 L}{2\kappa A^2\sigma} - \frac{\Delta T}{L}\right)x + T_H, \tag{1.62}$$

$$\frac{dT}{dx} = -\frac{I^2}{2\kappa A^2\sigma}x + \frac{I^2 L}{2\kappa A^2\sigma} - \frac{\Delta T}{L}. \tag{1.63}$$

At $x = 0$,

$$\left(\frac{dT}{dx}\right)_0 = \frac{I^2 L}{2\kappa A^2\sigma} - \frac{\Delta T}{L}. \tag{1.64}$$

Using Eq. (1.54) in Eq. (1.59), the expression for Q_{in} can be obtained as below,

$$Q_{in} = 2|S|T_H I + I^2 R_{leg} - 2\kappa A \frac{\Delta T}{L}, \tag{1.65}$$

where $R_{leg} = \frac{L}{\sigma A}$ is the intrinsic resistance of one leg.

In terms of Eqs. (1.54), (1.56), and (1.65), the conversion efficiency is given by

$$\eta = \frac{P}{Q_{in}} = \frac{\Delta T}{T_H} \frac{m}{\left[1 + m + \frac{R_{leg}}{R_{in}} \frac{\Delta T}{T_H} - \frac{(1+m)^2}{zT_H}\right]}, \tag{1.66}$$

where z is the figure of merit of the thermoelectric module,

$$z = \frac{2S^2}{\kappa R_{in}} \frac{L}{A}. \tag{1.67}$$

The internal resistance (R_{in}) includes the contact resistances and the resistances of the thermoelectric legs. When the contact resistances are neglected, z becomes the figure of merit (Z) of the materials,

$$Z = \frac{S^2 \sigma}{\kappa}. \tag{1.68}$$

In terms of the Eq. (1.66), the m value corresponding to the maximum efficiency is

$$m_\eta = \sqrt{1 + z\bar{T}} \tag{1.69}$$

where $\bar{T}$ is the mean operation temperature,

$$\bar{T} = \frac{1}{2}(T_H + T_C). \tag{1.70}$$

The maximum efficiency (η_{max}) is given by

$$\eta_{max} = \frac{\Delta T}{T_H} \frac{\sqrt{1 + z\bar{T}} - 1}{\sqrt{1 + z\bar{T}} + \frac{T_C}{T_H}}. \tag{1.71}$$

This equation is the same as the Eq. (1.32). This efficiency is always less than the Carnot efficiency (η_c),

$$\eta_c = \frac{\Delta T}{T_H}. \tag{1.72}$$

1.3 Peltier Cooling

The Peltier effect can be used for thermoelectric cooling. The Peltier cooling principle is quite different from the conventional cooling technologies that usually use compressed liquid or gas. Because of the compressor and circuit fluid, the conventional cooling instruments like air conditioners and refrigerators are quite bulky, heavy, and noisy. No moving part is required for a Peltier cooler; it is thus quiet and can have a very small size.

As shown in Figure 1.24, when an electrical current passes through the thermoelectric legs, both the electrons in the *n*-type leg and holes in the *p*-type leg transport from the cold side to the hot side. The charge carriers bring heat from the cold side to the hot side and thus cool down the cold side. The heat pumped at the cold side is $\Pi \times I$, where Π is the Peltier coefficient and I the current density. Π is related to the Seebeck coefficient by the Kelvin relationship, $\Pi = S \times T$.

Apart from the Peltier cooling, Joule heat is generated by the electrical current, and heat is transferred from the hot side to the cold side due to the temperature gradient. Half of the overall Joule heat transports to the cold side or the hot side. By neglecting the Thomson effect, the heat absorption at the cold side of a thermoelectric leg is given by

$$Q = ST_C I - \frac{1}{2} I^2 R - K\Delta T, \tag{1.73}$$

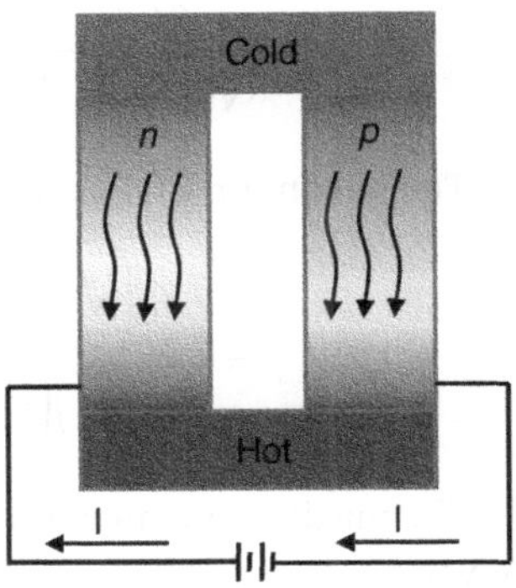

Figure 1.24 Schematic illustration of thermoelectric cooling. An external electricity source is connected to the two legs, and electrical current flows from the *n*-type leg to the *p*-type leg. The curved arrows indicate the transportation of the charge carriers from the cold side to the hot side in the thermoelectric legs.

where R is the electrical resistance of the thermoelectric leg, K the thermal conductance of the thermoelectric material, and $\Delta T = T_H - T_C$.

The input power by the electricity includes the heat pumping and the Joule heat,

$$P_{\text{in}} = S\Delta TI + I^2R. \tag{1.74}$$

The ratio of the heat absorbed at the cold side to the input electrical power is the energy efficiency of cooling. It is called as the coefficient of performance (COP),

$$\text{COP} = \frac{\text{heat absorbed}}{\text{electrical power input}} = \frac{ST_C I - \frac{1}{2}I^2R - K_{th}\Delta T}{S\Delta TI + I^2R}. \tag{1.75}$$

As shown in Eq. (1.73), the rate of heat pumping is a quadratic equation of the current I. There is a maximum rate of heat pumping. The corresponding current at the maximum heat pumping rate is given by

$$I_{\text{max}} = \frac{ST_c}{R}. \tag{1.76}$$

At this maximum cooling rate, the corresponding COP is given by

$$\text{COP}_{\text{max}} = \frac{\frac{1}{2}ZT_C^2 - \Delta T}{ZT_\text{H}T_\text{C}}. \tag{1.77}$$

The electrical current at the maximum cooling rate is different from that corresponds to the maximum COP. The electrical current corresponding to the maximum COP is given by

$$I'_{\text{max}} = \frac{S\Delta T}{R\sqrt{(1+Z\overline{T}} - 1}. \tag{1.78}$$

The maximum COP is shown by the Eq. (1.79),

$$\text{COP}'_{\text{max}} = \frac{T_\text{C}\left(\sqrt{1+Z\overline{T}} - \frac{T_\text{H}}{T_\text{C}}\right)}{(T_\text{H} - T_\text{C})\left(\sqrt{1+Z\overline{T}} + 1\right)}. \tag{1.79}$$

The maximum temperature difference between the hot and cold sides is given by

$$\left(T_{\mathrm{H}}-T_{\mathrm{C}}\right)=\frac{1}{2}ZT_C^2 \tag{1.80}$$

This indicates that thermoelectric materials with a high Z value are needed for efficient thermoelectric cooling. When the Z value is higher, lower temperature at the cold side can be achieved.

1.4 Thermoelectric Sensors

The Seebeck effect can be used to measure the temperature difference between two objects. One example is the thermocouples used for high temperature measurement. One end of a thermocouple is put to an object of constant temperature. The temperature of the other end can be read in terms of the Seebeck voltage.

1.5 Summary

Thermoelectric materials can be used for the direct energy conversion between heat and electricity. Apart from inorganic semiconductors and metals, intrinsically conducting polymers and organic molecules and composites of nanomaterials have also gained great attention. The ZT value is the parameter to characterize thermoelectric materials. A high ZT value requires high Seebeck coefficient, high electrical conductivity, but low thermal conductivity. Inorganic thermoelectric materials usually have a high Seebeck coefficient. But their thermal conductivity is very high. Organic or polymeric thermoelectric materials usually have a low thermal conductivity. But their Seebeck coefficient is lower than their inorganic counterparts by one to two orders of magnitude. The energy band structure and doping mechanism of conducting polymers is fundamentally different from metals or inorganic semiconductors. Thus, the approaches to improve the thermoelectric properties of conducting polymers can be quite different from that of inorganic thermoelectric materials as well. The thermoelectric properties of composites depend on the structure and properties of the fillers and the matrix and the microstructure of the composites. In-depth understanding of the thermoelectric fundamental knowledge is the key for the development of high-performance thermoelectric materials.

The applications of thermoelectric materials include thermoelectric generators, Peltier coolers, and thermoelectric sensors. There are heat transfer, drift electrical current, and thermoelectric effects during the operation of these devices. Their performance is thus strongly dependent of the physical properties of the thermoelectric materials, particularly the *ZT* value. Apart from the intrinsic properties of the thermoelectric materials, other factors such as the electrodes, the contact between the electrodes and thermoelectric materials and dimensions of the thermoelectric legs should be taken into consideration in designing a thermoelectric device.

Acknowledgment

This work was financially supported by a grant from the Ministry of Education, Singapore (R-284-000-228-112).

References

1 Bubnova, O. and Crispin, X. (2012). *Energy Environ. Sci. 5*: 9345.
2 Yao, H., Fan, Z., Cheng, H. et al. (2018). *Macromol. Rapid Commun. 39*: 1700727.
3 Fan, Z. and Ouyang, J. (2019). *Adv. Electron. Mater. 5*: 1800769.
4 Chen, Y., Zhao, Y., and Liang, Z. (2015). *Energy Environ. Sci.* **8**: 401.
5 Bubnova, O., Khan, Z.U., Malti, A. et al. (2011). *Nat. Mater. 10*: 429.
6 Li, Y., Ouyang, J., and Yang, J. (1995). *Synth. Met. 74*: 49.
7 Yao, H., Fan, Z., Li, P. et al. (2018). *J. Mater. Chem. A 6*: 24496.
8 Chiang, J.C. and MacDiarmid, A.G. (1986). *Synth. Met. 13*: 193.
9 Zhang, K., Qiu, J., and Wang, S. (2016). *Nanoscale 8*: 8033.
10 Song, H., Meng, Q., Lu, Y., and Cai, K. (2019). *Adv. Electron. Mater. 5*: 1800822.
11 Guan, X., Feng, W., Wang, X. et al. (2020). *ACS Appl. Mater. Interfaces 12*: 13013.
12 Guan, X., Yildirim, E., Fan, Z. et al. (2020). *J. Mater. Chem. A* 8: 13600 https://doi.org/10.1039/D0TA05324D
13 Guan, X., Cheng, H., and Ouyang, J. (2018). *J. Mater. Chem. A 6*: 19347.
14 Fan, Z., Du, D., Guan, X., and Ouyang, J. (2018). *Nano Energy 51*: 481.
15 MacDiarmid, A.G. and Epstein, A.J. (1995). *Synth. Met. 69*: 85.
16 Tigelaar, D.M., Lee, W., Bates, K.A. et al. (2002). *Chem. Mater. 14* (3): 1430.
17 Ouyang, J. (2013). *Displays 34*: 423.
18 Ouyang, J. (2018). *Acta Phys. -Chim. Sin. 34*: 1211.
19 Shi, H., Liu, C., Jiang, Q., and Xu, J. (2015). *Adv. Electron. Mater. 1*: 1500017.
20 Cao, Y., Smith, P., and Heeger, A.J. (1992). *Synth. Met. 48*: 91.

21 Xia, Y., Sun, K., and Ouyang, J. (2012). *Adv. Mater. 24*: 2436.

22 Ouyang, J. (2013). *ACS Appl. Mater. Interfaces 5*: 13082.

23 Mengistie, D.A., Ibrahem, M.A., Wang, P.C., and Chu, C.W. (2014). *ACS Appl. Mater. Interfaces 6*: 2292.

24 McCarthy, J.E., Hanley, C.A., Brennan, L.J. et al. (2014). *J. Mater. Chem. C 2*: 764.

25 Cao, Y., Smith, P., and Heeger, A.J. (1991). *Polymer 32*: 1210.

26 Pekker, S. and Janossy, A. (1981). *Mol. Cryst. Liq. Cryst. 77*: 159.

27 Lee, K., Cho, S., Park, S.H. et al. (2006). *Nature 441*: 65.

28 Tang, J., Chen, Y., McCuskey, S.R. et al. (2019). *Adv. Electron. Mater. 5*: 1800943.

29 Chen, G., Xu, W., and Zhu, D. (2017). *J. Mater. Chem. C 5*: 4350.

30 Chen, H., Ginzburg, V.V., Yang, J. et al. (2016). *Prog. Polym. Sci. 59*: 41.

31 Fanciulli, C., Abedi, H., Merotto, L. et al. (2018). *Appl. Energy 215*: 300.

32 Menon, A.K. and Yee, S.K. (2016). *J. Appl. Phys. 119*: 055501.

33 Aranguren, P., Roch, A., Stepien, L. et al. (2016). *Appl. Therm. Eng. 102*: 402.

34 Madan, D., Wang, Z., Chen, A. et al. (2013). *ACS Appl. Mater. Interfaces 5*: 11872.

2

Conductive Polymers for Flexible Thermoelectric Systems

Lin Hu[1], Zaifang Li[1], Yinhua Zhou[2], and Fengling Zhang[3]

[1] *China-Australia Institute for Advanced Materials and Manufacturing, Jiaxing University, Jiaxing, P. R. China*
[2] *Wuhan National Laboratory for Optoelectronics, and School of Optical and Electronic Information, Huazhong University of Science and Technology, Wuhan, Hubei, China*
[3] *Department of Physics, Chemistry and Biology, Linköping University, Linköping, Sweden*

2.1 Introduction

Energy crisis has become one of the major challenges with the acceleration of industrialization and population growth in the world. Researchers are searching for alternative energy sources (such as sunlight, wind, terrestrial heat, etc.) and developing various energy conversion technologies to meet the societal energy demands. Thermal energy is abundant but often wasted in daily life and industrial production. Thermoelectric generators or systems, which can directly convert the waste thermal energy into useful electrical energy, are promising and efficient technologies in solving the increasingly serious energy crisis. In recent years, organic thermoelectric materials, especially the conductive polymers, have attracted tremendous attention and been intensively studied owning to the advantages such as light weight, excellent mechanical flexibility, easy processability, and suitability for flexible and wearable applications [1–3].

Conductive polymers, more precisely, intrinsically are polymers with an extended electron delocalized system through the alternating single (σ) and double (π) bonds. These materials have metallic conductivity or can be semiconductors due to the presence of a conjugated π-electron backbone. Their physical, chemical, and electronic properties can be fine-tuned using the methods of organic synthesis or controllable doping techniques. Moreover,

Flexible Thermoelectric Polymers and Systems, First Edition. Edited by Jianyong Ouyang.

like other common organic materials, the conductive polymers can also be solution-processed (by spin-coating, printing, slot-die coating, or others) from their solutions or dispersion for applications. Therefore, they have been extensively employed in the devices like thin-film transistors [4], organic light-emitting diodes [5], photovoltaic cells [6], and thermoelectric generators [7]. In this chapter, we will focus on the applications of conductive polymers in thermoelectric devices or systems. Conductive polymers intrinsically have low thermal conductivity and high Seebeck coefficients, rendering them potential candidates for high-performance thermoelectric applications [8, 9]. This chapter will introduce the fundamental knowledge of conductive polymers, including the development of conductive polymers, chemical structures and synthesis of the polymers, doping techniques, representative poly(3,4-ethylenedioxythiophene) (PEDOT), and processing for flexible thermoelectric generators.

2.1.1 The Discovery and Development of Conductive Polymers

Polymers had long been regarded and applied as insulators until the discovery of electrically conducting polyacetylene (PAc) in 1977 by Professors A. J. Heeger, Alan G. MacDiarmid, and H. Shirakawa. They obtained a high conductivity of 38 ohm^{-1} cm^{-1} in oxidized iodine-doped PAc [10, 11]. This discovery attracted scientists' tremendous attention and initiated the field of conductive polymers, although the PAc itself was not found in practical applications. Since then, a huge and growing number of scientific publications, promising processing techniques and even various commercial uses of the conductive polymers, have been reported [12]. Three scientists were awarded the Nobel Prize in Chemistry in the year 2000 "for the discovery and development of conductive polymers."

In reality, the reports on conductive polymers could be traced back to the mid-nineteenth century. Henry Letheby, a chemistry professor at the College of the London Hospital, first obtained polyaniline (PAni) through the electrochemical and chemical oxidation of aniline in acidic media [13]. Although the nature of the products and its potential remained unidentified at that time, it was the earliest landmark in electropolymerization. The first highly conductive organic compounds were charge transfer complexes. In 1954, M. Yoshio and coworkers reported that the organic charge transfer complexes between polycyclic aromatic compounds and halogens delivered a fairly high conductivity ($1\sim10^{-3}$ ohm^{-1} cm^{-1}) [14]. And later, researchers also demonstrated that the complex salts of tetrathiafulvalene (TTF) even showed almost metallic conductivity [15]. It indicated that organic compounds could carry current, though these compounds were not polymers. In 1963, three Australian scientists B. A. Bolto, R. McNeill, and D. E. Weiss demonstrated the derivatives of polypyrrole (PPy) with conductivity as low as 1 S cm^{-1} [16]. They concluded that the polymer may behave as an intrinsic or extrinsic semiconductor with n- or p-type characteristics. Subsequently, R. De Surville and coworkers also reported high conductivity in PAni [17]. Moreover, Diaz and Logan further reported films of PAni that can serve as electrodes [18]. After the spectacular progress for PAc was published in 1977, this new field of conductive polymer was born.

2.1.2 Representative Structures

Electrically conducting polymers can be divided into three main groups: polyenes, polyheterocycles, and polyaminoaromatic compounds. PAc, PPy, PAni, polythiophene (PTh), and its derivatives are the most intensively studied intrinsically conductive polymers. The representative structures are shown in Figure 2.1a.

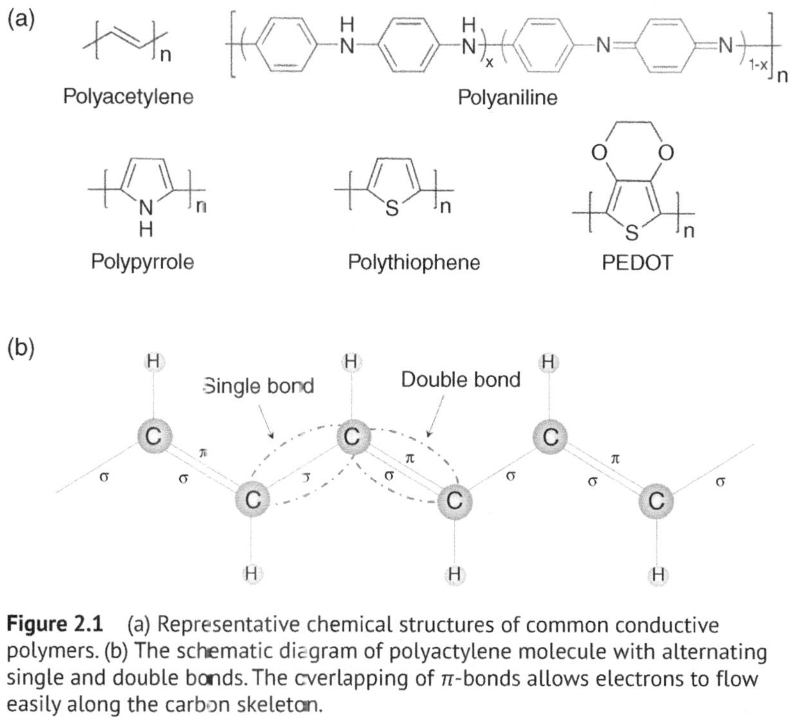

Figure 2.1 (a) Representative chemical structures of common conductive polymers. (b) The schematic diagram of polyactylene molecule with alternating single and double bonds. The overlapping of π-bonds allows electrons to flow easily along the carbon skeleton.

2.1.2.1 Polyacetylene (PAc)

PAc is one of the simplest conjugated polymers with a repeating unit $(C_2H_2)_n$. The name is derived from its conceptual construction through polymerization of acetylene to give a chain with repeating olefin groups. Due to the non-twistable nature of the double bond, each repeating unit has both cis and trans structures. PAc thus also exists in various isomeric forms. The cis-PAc or trans-PAc can be controlled by changing the temperature and catalytic agents during the syntheses [19]. It is worth noting that cis-PAc delivers much lower conductivity than those of the trans isomers since the cis structure causes the distorted molecular chains and the π-electron delocalization is hindered [20]. Moreover, the cis-PAc is relatively thermodynamically unstable. In addition, early work indicated that the PAc products always presented in a linear type with high crystallinity and were insoluble, infusible, and air sensitive [21, 22]. A major development of PAc was made by H. Shirakawa's group. They obtained silvery films of PAc via preparing the polymerization at solution surface of the catalyst system [23]. Furthermore, collaborated with A. J. Heeger and Alan G. MacDiarmid they found that the polymer displayed tremendous enhancement in conductivity up to seven orders of magnitude upon doping with iodine [11]. Similar results were also achieved using halogens chlorine, bromine, and arsenic pentafluoride (AsF_5) [24]. It was found that the conductivity of cis-PAc doped with AsF_5 could be closed to that of copper [25]. Up to now, the conductivity of PAc can be controlled ranging from insulator to semiconductor to metal with the improved synthesis conditions and doping techniques. Despite the discovery of PAc as a conductive polymer leading to many developments in materials science, there are almost no commercial applications for this polymer. The PAc still suffers from many drawbacks such as instability to air and difficulty in processing, etc. Therefore, attention has shifted to many other conductive polymers for applications.

2.1.2.2 Polyaniline (PAni)

The discovery of PAni has been more than 150 years [13]. The compound itself did not capture much attention in scientific community until the successful synthesis and doping of PAc in 1977. In the past decades, PAni has been of particular interest because of its low cost, high environmental stability, solution- or counterion-induced processability, and controllable electrical conductivity [26]. Especially, the electrical properties switch between its conductive and resistive states by an appropriate doping/dedoping process [27]. The synthetic methods of PAni are quite simple. PAni is usually prepared by direct polymerization of aniline through chemical or electrochemical oxidation. Therefore, the polymer always exists in a variety of forms that differ in chemical and physical properties based on its oxidation level. The general formula of PAni contains a reduced repeat unit and an oxidized repeat unit, as shown in Figure 2.1a. In the figure, the structures are poly (p-phenylene amine imines) when $0<x<1$ and the oxidation state of the polymer increases with increasing content of the imine form. The fully oxidized state of PAni is pernigraniline ($x = 0$), the halfoxidized state is emeraldine ($x = 0.5$), and the fully reduced state is leucoemeraldine ($x = 1$) [28]. Each structure can exist in the form of its base or salt by treatment of the base with an acid. Studies have shown that most forms of PAni are one of the three states or physical mixtures of these components. Note that, the emeraldine is the most stable and neutral. The imine nitrogens can easily be protonated by an acid (doped) to form its salt. Research shows that the protonation helps to delocalize the otherwise trapped diiminoquinone–diaminobenzene state and greatly improves the conductivity [29].

The excellent processability, together with the presence of a number of intrinsic redox states, enhanced the potential applications of PAni in practical devices. The color change associated with PAni in different oxidation states can be used in sensors and electrochromic devices [30]. Moreover, the electrical conductivity can easily be regulated between different oxidation states or doping levels, which can also be utilized to improve the response characteristics of biosensors [31]. In addition, PAni is available as powder, film, or fibrils, further expanding its applications. It has been widely employed as surface treatment, antistatic coating, and corrosion protection in printed circuit board manufacturing [32]. PAni is also primarily of interest as an electrode material in the lithium ion batteries because of its high discharge capacity [33].

2.1.2.3 Polypyrrole (PPy)

The formula of PPy is $H(C_4H_2NH)_nH$. PPy is described as "quasi unidimensional" polymer due to the existence of cross linking and chain hopping [34]. Although PPy is not reported as early as PAni, it emerged as an electrically conductive polymer earlier than PAni. It is the first example of polyheterocycles as conductive polymer. PPy can also be prepared through chemical or electrochemical oxidative polymerization of the pyrrole monomers. However, the processing and applications of the synthetically conductive PPy was restricted by its insolubility and infusibility. Further investigations on the compounds have also been retarded. Many researchers thus pay their attention to develop soluble or swollen PPy derivatives, such as poly(3-alkylpyrrole) [35], poly(1-(2-carboxyethyl)pyrrole) [36], poly(N-substituted pyrrole) [37], etc.

PPy itself is an insulator, while it converts into a conductive polymer upon an oxidative doping. The conductivity depends on the conditions and reagents used in the oxidation. Particularly, the conducting anion in the reagents would be incorporated into the polymer as a counterion during the doping process [38]. Therefore, properties of the counterion (e.g., its size, geometry, and charge) would influence the electrical properties of the polymer. A higher conductivity is always associated with larger anions, such as tosylate. In addition, a pristine PPy film (undoped) is yellow and turns dark in air due to some oxidation. The color of the doped films (blue or black) is also related to the degree of the oxidative doping as well as film thickness. Doped PPy has been one of the most thoroughly investigated conductive polymers because of the excellent properties including electrical, thermal, and mechanical properties. It has been widely investigated and employed in many microelectronics fields [39]. PPy was used as the polymeric electrodes for rechargeable lithium batteries due to the advantages of light weight combined with high energy storage capacity and a possibility to manufacture them in thin films [40]. The electrical conductivity of PPy is sensitive to chemical agents, endowing its potential use as a gas sensor [41]. The electropolymerized free-standing PPy films are also good candidates for thermoelectric applications [42].

2.1.2.4 Polythiophene (PTh) and Derivatives

PTh with a formula $(C_4H_2S)_n$ is polymerized from the sulfur heterocycle, thiophenes. The thiophene monomers are linked through 2- and 5-positions of the rings. It is a white solid with poor processing ability, while the remaining 3- or 4-position in the rings can always be substituted by functional group to improve the solubility of the derivatives. Note that, the thiophene was considered to be an attractive monomer for conducting polymers after decades of the first report on thiophene polymerization [43]. Interest is still growing because thiophene is easier to handle (less sensitive to oxygen and moisture) and allows simple preparations through substituted monomers to adjust the properties (solubility, processibility, and optical and electrical characteristic) of the conjugated polymers.

PTh and its derivatives take on a dark color and become conductive when oxidized. But the thiophene monomers need stronger oxidants for the polymerization than pyrrole or other heterocycles. A variety of reagents have been used to dope the polymers, including halogen, organic acids, and ferric chloride [44–46]. In principle, the electrical conductivity results from the delocalization of electrons along the extended π-systems of the polymer backbone. Meanwhile, the overlap of π-orbitals of the aromatic rings also produces some interesting optical characteristics of these materials. Usually, the absorption of the polymers redshifts with the increase of the conjugated fused rings. Changes in both color and conductivity are closely related to the conjugation length and twist of the polymer backbone [47]. Therefore, one of the most feasible strategies to get desired properties of the polymer is to functionalize the thiophene monomer with different groups. Soluble polymers are derivable from 3-position substituted of thiophene rings where the substituent is butyl or longer. The polymer poly(3-alkylthiophene) (P3HT), a milestone in the development of PTh derivatives, was found to be processed into film and possessed excellent photoelectric properties [48]. Since then, many derivatives of PTh have been synthesized, characterized, and widely employed for applications (display devices, energy storage, solar cells, and biotechnology) in terms of the conductivity, processability, and stability [49–51]. Especially, the properties and applications of PEDOT, one of the most successful PTh derivatives, would be discussed in detail in the fourth section.

2.1.3 Conductive Mechanism

Understanding the charge transport mechanism in the conductive polymers is essential for basic insight of their thermoelectric properties. It is well known that the electrical conductivity of a material is mainly determined by its electronic structure. The energy band theory accounts for the different behaviors of metals, semiconductors, and insulators. The bandgap is the energy difference between the highest occupied energy level (valence band) and the lowest unoccupied energy level (conduction band) of the material. Metals have a negligible bandgap. The electrons can move freely, resulting in a high electron mobility and conductivity. Semiconductors possess a narrow bandgap that electrons can cross from the valence band to the conduction band upon excitation. The electrons excited in conduction band and the resulting holes left behind in valence band contribute the charge transport. If the bandgap of the material is large enough (>3 eV), electrons hardly cross the gap, which makes the material insulating. In traditional insulating polymers such as polyethylene, the valence electrons are bound in sp3 hybridized covalent bonds to form sigma bonds (σ-bonds). Such shared "σ-bonding electrons" do not contribute to the electrical conductivity of the material. However, the situation is completely different for the conjugated polymers. Take PAc, a conjugated polymer with the simplest structure, as an example. As shown in Figure 2.1b, it exists contiguous with sp2 hybridized carbons as the backbone of the polymer. The three electrons in the hybrid orbits form three sigma-bonds. On the other hand, one valence electron on each carbon atom resides in a p_z orbital. All the p_z orbitals, which is orthogonal to the σ-bonds, combine with each other to form a molecule wide delocalized set of orbitals (π-bonds). The overlapping of π-bonds results in facile movements of the delocalized electrons along the polymer backbone [8]. Based on the Peierls instability, the π-band is divided into bonding orbital π (filled) and anti-bonding orbital π^* (empty) with an opening of a gap in the electronic excitation spectrum [52]. This band-like energy distribution among electronic density of states in the conjugated polymers resembles the traditional semiconductors. The energy gap E_g is the energy spacing between highest occupied molecular orbital (HOMO) and lowest unoccupied molecular orbital (LUMO) of the polymers. The intrinsically conjugated polymers have poor electrical conductivity. However, electrical conductivity of polymers can be enhanced by doping or other excitation methods [53, 54]. These methods can help generate charge vacancies in the energy band of the polymer. An electron at a neighboring position would jump and fill the vacancy, generating a new vacancy and allowing charge to flow along the polymer chains.

2.2 Chemical Design and Synthesis of Conductive Polymers

Up to now, there is only a few well-known conjugated polymers having been exploited and studied in organic thermoelectrics [55]. However, it can be anticipated that the recent continuously increased interest in flexible and wearable energy harvesting applications will stimulate a surge development of specifically designed novel thermoelectric polymers. The unique advantages of the conductive polymers are enabling the rational design and synthesis of molecular structures, tunable band structures, and controllable doping techniques. Aside from these advantages, factors such as cost, processability, and stability of the monomers and polymers should also be taken into account in a practical and commercial thermoelectrics technology. Therefore, a fundamental understanding of the chemical design and synthesis of the versatile conductive polymers allows for an effective tailoring of their intrinsic properties to meet the application needs.

2.2.1 Energy Level Design of Conjugated Polymers

The conductive polymers have alternating single and double bonds along the polymer skeletons, rending them with unique semiconducting properties. Chemical coupling of the conjugated monomers into long polymer chains leads to orbital interaction and energy level splitting. The magnitude of the energy gap E_g and the positions of the HOMO and LUMO energy levels are the most important characteristics to determine the electrical properties of a given conjugated polymer [56].

Usually, there are two possible resonant structures: benzenoid and quinoid forms, for the conjugated polymers at ground states with nondegenerate energy. Such as polyphenylene and PTh, each benzene or thiophene unit maintains its aromaticity with confined π-electrons in the benzenoid form. While delocalization of the π-electrons along the conjugated chains causes the transformation between the single and double bonds, leading to a resonant structure referred to as the quinoid form. Compared to the benzenoid form, the quinoid form increases the planarity of the polymer backbones and enables more efficient π delocalization along the polymer chains. The conjugated polymers thus display a smaller HOMO–LUMO bandgap [57]. A certain polymer naturally prefers to adopt the quinoid structure in the conjugated backbone. It is apparent that adopting a quinoid structure requires destruction of the pristine aromaticity. A reduction of the aromaticity in the conjugated main chains allows a greater tendency to the quinoid form through π-electron delocalization. A typical example is polyphenylene ($E_g = 3.2\,eV$) and poly(phenylenevinylene) ($E_g = 2.4\,eV$) [58]. The bandgap is reduced by inserting a double bond to dilute the effect of the benzene rings and reduce the aromaticity. The same effect occurs in the polyisothianaphthene, which has very narrow bandgap of 1.1 eV [59].

A more effective strategy in tuning the energy levels of the conjugated polymers is to design and synthesize donor–acceptor (D-A) alternating copolymers [60]. These copolymers incorporate a conjugated electron-rich donor (D) unit and a conjugated electron-deficient acceptor (A) unit in the same polymer backbone. The electron delocalization and the formation of quinoid structures (-D-A- to -D^+-A^--) are facilitated owning to the push-pull effect caused by the D and A units. One unique feature of these polymers is that the energy levels as well as the bandgap of the polymers can be tuned by engineering the donor and acceptor units separately. Empirically, their HOMO and LUMO energy levels are dominated by the HOMO of the donor segment and LUMO of the acceptor segment, respectively [61]. Figure 2.2 summarizes a series of structure fragments employed to construct the D-A alternating copolymers. A judicial choice of the donor and acceptor monomers can tune the energy levels of the copolymers to a desired magnitude.

In addition, molecular modification used to impose steric or electronic effects on the conjugated main chains is also useful strategy to adjust the energy level and bandgap of the polymers. Planarization between adjacent aromatic units allows parallel p-orbital interactions to extend conjugation, facilitate π-electron delocalization and reduce the bandgap of the

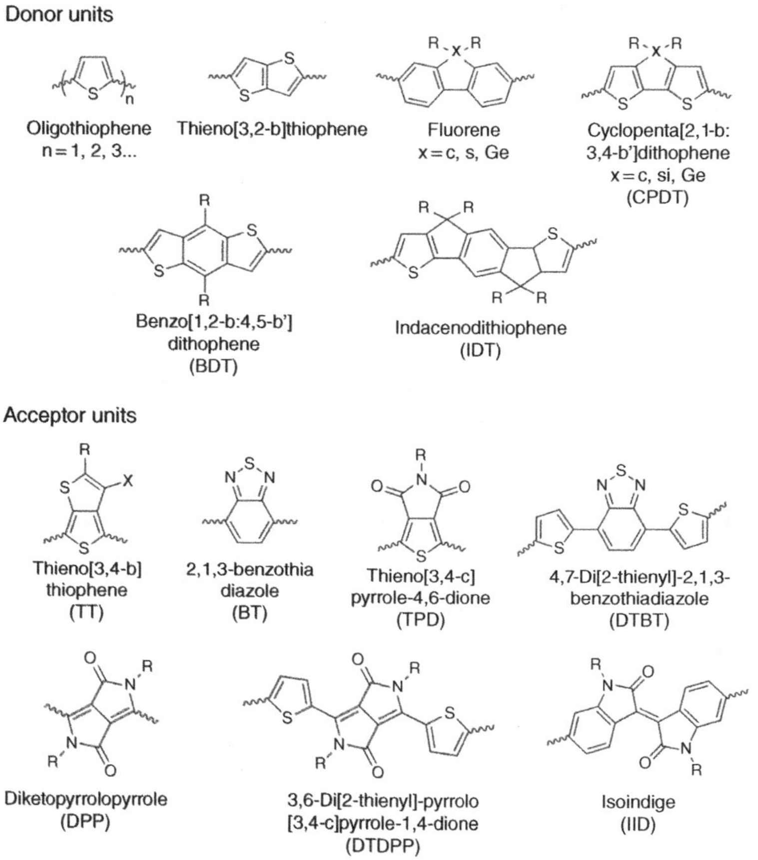

Figure 2.2 The structures of some widely used donor and acceptor fragments.

conjugated polymers. However, problems may occur when the pendent conjugation is too large. The extra introduced steric hindrance and torsion would inhibit the molecule with a quinoid structure in the backbone. Moreover, it is also detrimental to the intermolecular π-π stacking interaction as well as the solubility and processability of the polymers [56]. Apart from the conjugated side-chain modification, incorporation of electron-donating or electron-withdrawing substituents directly onto the polymer backbone is another effective way of perturbing the molecular orbitals. The electronic properties of the resulting polymers, such as energy levels and dipole moments, can be strongly influenced. In general, the electron-donating groups (fluorine, cyano group, etc.) raise the HOMO energy levels, while electron-withdrawing groups (amino, alkoxy, sulfydryl group, etc.) lower the LUMO energy levels of the polymers. A represent example is: the bandgap of the well-known conductive polymer PEDOT, with direct attachment of alkoxy groups, is about 0.5 eV lower than that of its parent PTh [62].

2.2.2 Tuning Molecular Conformations

The electrical properties of conductive polymers and their performance in thermoelectrics application are intimately tied to the molecular conformations and microstructures in the solid states. It is well known that conjugated polymers are weakly bound macromolecules with many conformational degrees of freedom. Their crystallites can either be randomly oriented or display preferential orientation to each other. The chain conformations would not only affect the phase behavior (amorphous to crystalline), but also make a difference to the aggregate domains in both liquids and films [63]. As referred to the inorganic semiconductors, charge transport is superior of the polycrystalline analogues than that of their amorphous. Much of progress in the design of the conjugated polymers was made by pursuing polycrystalline-like microstructures. Typical crystalline packing involves ordered lamellae composed of co-facially stacked molecules that exhibit strong π-orbital overlap. Li and co-workers firstly put forward two-dimensional (2D) conjugated PThs. They found that the 2D structure effectively enhanced the hole mobilities and absorption spectra of the PThs [64]. Considering that the intermolecular charge transport occurs in a hopping model through the overlapped π-electron orbits between the adjacent conjugated backbones, they further developed a series of derivatives by introducing various conjugated side groups to extend the electron delocalization [65]. It is worth noting that the molecular conformations as well as the stacking strength and mode also significantly affect the solubility, film morphology, and even kinetic conditions of the film-processing methods [66]. Thus, a synergetic balance is required between extending the conjugation system to enhance the charge-transport properties and tailoring the non-covalent side-chain interactions to maintain proper processability. In addition, to attain excellent electrical properties, recent polymers [67–69] based on the increasingly complex monomers broke the old paradigms, such as involving regioregularity, texture, or crystallinity. Some highly disordered or even seemingly amorphous polymers even possess a comparable conductivity to the semicrystalline materials [70, 71].

2.2.3 Melt and Solution Processability

The target conjugated polymers should be able to be melted and possess a reasonable solubility in common organic solvents to ensure their processability. As discussed in Section 2.2, the unsubstituted conjugated polymers such as PPy, PAni, and PTh are intractable, due to their poor insolubility and infusibility. These polymers can be either processed in the form of a solid powder or deposited in situ as a thin film on an electrode through electrochemical polymerization, but the subsequent doping and applications of the polymers are seriously limited.

A feasible strategy to solve the problems is the formation of a soluble polymer:dopant complex during the oxidative polymerization process. Counterions with an excellent solubility are always introduced to induce this processability. The extensively studied complex of PEDOT and the counterion PSS, is synthesized through oxidative polymerization of 3,4-ethylenedioxythiophene in the presence of PSS in solution [72]. Sodium persulfate is usually used as the oxidizing agent. The obtained PEDOT:PSS complex can be dispersed uniformly in water, which is an excellent choice of solvent. Another widely used approach to enhance the processability of conjugated polymers, especially in solution methods, is the decoration with alkyl side chains. It is well known that the solubility of a given polymer is governed by several structural factors, including the backbone rigidity, the degree of polymerization, polymer regioregularity, and the chain lengths of pendent aliphatic groups [73, 74]. Strong interchain π-π stacking interaction is considered as the directly contributing reason leading to the conjugated polymers with poor insolubility. The pendent aliphatic groups play an important role in determining the intermolecular interactions between different polymer chains and influencing the solubility. In general, large and branched alkyl chains are more effective to improve the solubility of polymers in common organic solvents. However, the reduced π-π interchain interaction by the insulating alkyl chains may result in deterioration in the charge-transport. Therefore, the sizes and shapes of the alkyl chains as well as their positions need to be judiciously selected for designing the conjugated polymers with desired properties. Bao and co-workers have reviewed the side-chain engineering in solution-processable conjugated polymers for organic solar cells and organic field-effect transistors [75].

2.3 Doping of Conductive Polymers

Doping refers to the introduction of impurities into the intrinsic semiconductor to adjust its electrical, optical, and structural properties. It can be divided into n- and p-type doping according to which apparent carrier mobility (electrons or holes) is largest in the doped system. Owing to the unique chemical structures, the doping mechanism for conjugated polymers is completely different to that of their inorganic counterparts. The doping processes are mainly realized by adding chemical reactants to oxidize or reduce the polymers. In general, a favorable energy alignment is required between the conjugated polymers and the dopants for an efficient doping, as shown in Figure 2.3a [76]. Electrons are redistributed or transferred to create charge carriers in the form of polarons (radical ions), bipolarons (dications or dianions), or solitons in the conjugated system [77]. The charge carrier concentration as well as the electrical conductivity depends on the doping level, which can be modulated through the dopant concentrations. The movement of these charge carriers along polymer chains produces conductivity. It is well known that neutral or undoped conjugated polymers have always been reported as insulators or semiconductors, while their conductivity can be increased to metallic level upon doping.

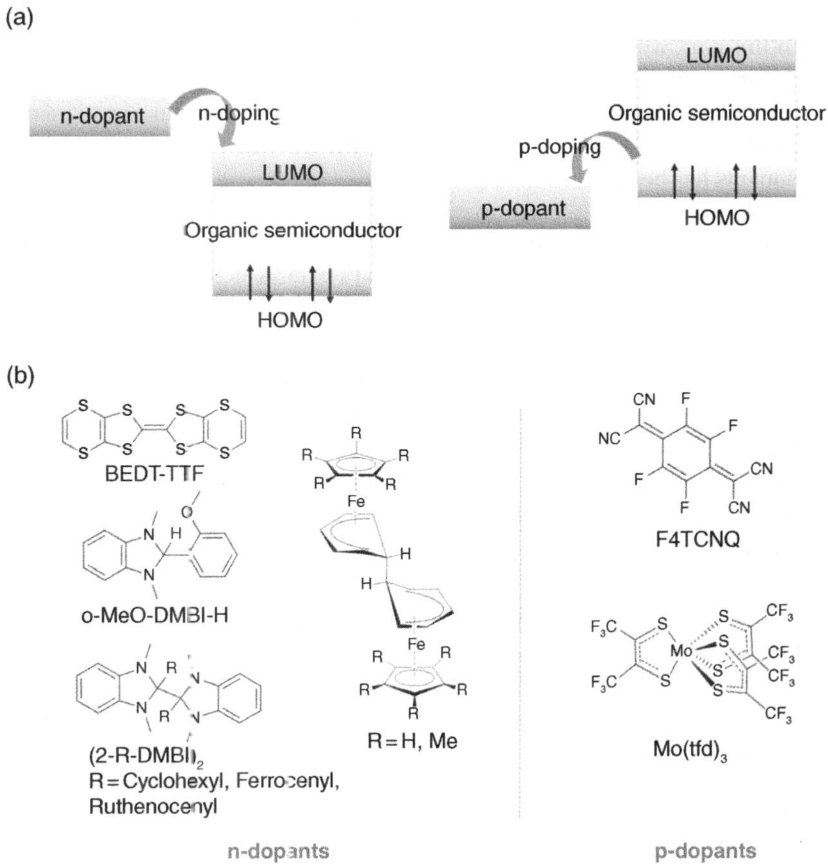

Figure 2.3 (a) Mechanism of the organic semiconductor doping process. (b) Representative chemical structures of common dopants for conductive polymers.

2.3.1 n-Type Doping

It can be seen from Figure 2.3a that n-type dopant can be regarded as an electron donor during the doping process, donating electrons to the organic semiconductor. To make the electron transfer efficient, the ionization energy of the dopant has to be lower than the electron affinity of the host materials. Thus, n-type doping is intrinsically more challenging due to an easy oxidization of the dopants by ambient oxygen and water [78]. Moreover, the electron-rich, n-doped polymer can also react with the elemental oxygen to de-dope the polymer. These factors require that n-type doping be performed in an environment of inert gas. The development of stable n-type dopants is an urgent problem to be solved. Accordingly, much of the work to develop new n-type dopants has been focused on enhancing the air-stability of either the precursor dopants or the doped charge transfer states. In this part, we will briefly review and compare the major classes of n-type dopants (alkali metals, organic radicals, organometallic compounds and low ionization potential donor molecules) commonly used in organic semiconductors and their applications in organic electronics. A series of n-type dopants has been shown in Figure 2.3b.

Alkali and alkaline earth metals (Li, Na, K, etc.) are the first and most fundamental class of n-type dopants. They can reduce a wide range organic semiconductors even some materials with LUMO levels above −3 eV [79–81]. However, the alkali cations produced after the reduction diffuse readily through organic layers, leading to instabilities in the doping systems. Moreover, it is not only inconvenient to operate but also dangerous to handle due to the fact that alkali and alkaline earth metals are active metals. A much more easily handled class of n-type dopants is the corresponding oxy metal salts. It has been reported that lithium carbonate can effectively dope conjugated polymers to improve the conductivity as electron transfer layer for highly efficient polymer light-emitting diodes [82]. Note that, these salts are processed via high temperature vacuum deposition and the doping mechanism is not clear. Then Karl Leo and co-workers found some organic molecules with low ionization energies could also be used as n type dopants. They firstly reported n-type doping of the naphthalenetetracarboxylic dianhydride (NTCDA) by co-evaporating the NTCDA molecule with the donor molecule bis(ethylenedithio)-tetrathiafulvalene (BEDT-TTF) in year 2000 [83]. The conductivity of the doped NTCDA film was one to two orders of magnitude higher than that of the pristine films. This indicated that n-type doping of the organic semiconductor can be achieved by the organic molecular compounds. Moreover, compared with the abovementioned dopants such as an alkali metal and an ionic salt compounds, the doping system is better in thermal stability for the avoidance of ion diffusion. The scientific community began to explore this class of materials. More organic molecular compounds were developed to be effective n-type dopants in both polymers and small molecules [84–86]. These n-type dopants are air sensitive and have to be performed with extra care under inert atmosphere due to the materials with an extremely low ionization energy. More recently, new classes of stable n-type dopants, such as hydrides and dimers, by releasing strongly reducing species in situ from their precursors have emerged. The representative structures 2,3-dihydro-1H-benzimidazoles (DMBI-H) and its derivatives have been widely used as n-type dopants for [6,6]-phenyl-C_{61}-butyric acid methyl ester ($PC_{61}BM$) and poly[*N*,*N*′-bis(2-octyl-dodecyl)-1,4,5,8-napthalenedicarboximide-2,6-diyl]-alt-5,5′-(2,2′-bithiophene) (P(NDI2ODT2) in organic electronics [87, 88]. It was indicated that the doping process begins with a hydride transfer and electron transfer steps result in the formation of the host radical anions subsequently. The organic dimer compounds as n-type dopants were firstly reported in a patent [89]. The concept of dimers as n-type dopants is based on the desire to stabilize some air sensitive dopant radicals by dimerization through a weak C-C σ bond. Radicals that are easily oxidized can be stabilized in the air for the requirement of a small amount of energetic cost to break the C-C bond. An outstanding advantage of these dopants is their good air stability. Moreover, they can be processed in both solution and vacuum conditions. The size of the cations produced in the doping process can be controlled by manipulating the ion diffusion rate, the evaporation rate, or the morphology of the films. Organometallic dimers of 19-electron metallocenes were first reported as n-type dopants by Seth Marder and Antoine Kahn [90]. This work was further extended to a series of DMBI-based organic radical dimers. A comparison of the doping effect of $(DMBI)_2$ and aforementioned DMBI-H revealed faster formation of the host anions and a more reliable and stable doping effect between the dimer and the hosts, although the two dopants possess similar electron affinities [91]. Due to the importance of ntype doping for organic electronics, a wide range of other categories of ntype dopants have also been carried out to dope the conjugate polymers, including self-doped anions, self-compensated anions, amine, and amidine derivatives [92–95].

2.3.2 p-Type Doping

In contrast to n-type dopants, p-type dopants can be viewed as electron acceptors during the doping process. As shown in Figure 2.3a, electrons will transfer from the organic semiconductor to the dopant and holes are formed in the organic semiconductors simultaneously. To facilitate the electron transfer, the electron affinity of the p-type dopant is usually higher than the ionization energy of host material. Since the p-type dopant has a relatively high energy level, it is also more stable in air and easy to prepare. The p-type doping is thus more popular and reliable for practical applications. According to literature, p-type dopants can be classified into several categories including diatomics, metal oxides, Bronsted acids and Lewis acids, and small molecules, etc. Some structures are also presented in Figure 2.3b.

The diatomics (I_2, Br_2, Cl_2, and O_2) is the earliest examples for p-type doping. As mentioned previously, the PAc could become conductive when exposed to iodine vapors through a redox reaction [12]. Since then many conjugated polymers characterized by a tunable electrical conductivity have been designed and synthesized. In addition to doping PAc, halogens or oxygen was also used to dope PPy, PThs, and other conjugated polymers [96]. Metal oxide p-dopants mainly refer to some transition metal oxides with high valence, such as MoO_3, V_2O_5, WO_3, etc. [97–99]. The electron affinities of these metal oxide dopants are closely related to the processing conditions. MoO_3 is the most common metal oxide p-type dopant because it can be evaporated at relatively low temperatures (~400 °C) and then deposited into a film. Other metal oxides, such as V_2O_5 and WO_3, require much higher temperatures to be vaporized, which may result in the loss of oxygen in the molecule. Many research groups have studied the preparation of metal oxides by sol–gel methods to avoid the use of such high-temperature deposition techniques. However, different preparation methods lead to the formation of diverse structures and particle sizes of metal oxide clusters, which, in turn, will affect the doping efficiency of organic semiconductors [100]. Practically, Bronsted acids and Lewis acids are the most widely used p-type dopants for the doping of conjugated polymers. The transfer of a proton (H^+) to the backbone of the conjugated polymer contributes to the p-type doping. A prominent example is PAni. The emeraldine, one structure of PAni, turned into highly conductive emeraldine salt upon protonation, with an electrical conductivity of more than $100\,S\,cm^{-1}$ [101]. Another widely studied, commercially available, and stably doped ptype material poly(3,4-ethylenedioxythiophene) (PEDOT) also belongs this type of doping, which will be described in details in Section 2.4. Some small molecules with strong electron-withdrawing ability such as F4TCNQ and other related TCNQ compounds can also be used as p-type dopants. F4TCNQ is a strong electron acceptor with an electron affinity of 5.2 eV. And its planarity enables blending with many hosts including both small molecules and polymers, to induce efficient p-type doping [102].

2.4 The Properties of Poly(3,4-ethylenedioxythiophene)

2.4.1 Oxidative and in situ Polymerization of EDOT to PEDOT

PEDOT, derivative of PTh, is the most widely used conductive polymer due to its high electrical conductivity, good physical and chemical stability, superior optical transparency and the capabilities of easy doping and solution processing [103]. The synthesis of PEDOT by chemical polymerization is usually according to two principal methods, i.e. oxidative and in situ polymerization. In both cases, a conducting material is obtained since the oxidant can simultaneously p-dope the conjugated polymer PEDOT during the polymerization

The first systematic experiments for oxidative polymerization of EDOT to PEDOT referred to the synthesis routes for PPy. Iron (III)-chloride was used as the oxidant in the reaction and insoluble solid powders of the polymer with a tremendously high conductivity were obtained [104]. It was also found that the conductivity is related to the temperature of the solvents, indicating an enormous temperature stability of PEDOT [105]. Then, several other metal ions, such as Cerium (IV) [106], Manganese (IV) [107], and Copper (II) [108], have been used with varied success to oxidize EDOT. However, a technically useful breakthrough is the preparation of highly conductive PEDOT layers in lieu of the intractable solid powders by using iron (III)-sulfonates as oxidants [109]. The sulfonates are soluble in common organic solvents like ethanol or n-butanol which would not be oxidized by Fe (III). For example, PEDOT dispersions could be obtained through mixing EDOT and Fe (III)-tosylate in alcoholic solution without immediate reaction or forming undesired precipitates. Note that the pot life of these mixtures is limited. While it was found that the pot life can be enhanced as well as the conductivity by addition of organic bases such as imidazole or pyridine into the reaction mixture as a basic inhibitor [110, 111]. These observations were attributed to the formation of toluenesulfonic acid during the polymerization of EDOT. Toluenesulfonic acid and its anion not only function as the counterions for the positively charged (doped) EDOT moieties in the polymer, but also are byproducts of the oxidative polymerization. The acidity during the polymerization process was controlled through these basic inhibitors. Besides iron (III) salts, some special peroxides like hydrogen peroxide, alkyl hydroperoxides, and diacyl peroxides, are also able to polymerize EDOT and subsequently dope PEDOT to yield highly conductive PEDOT cations. In addition, an advanced process of vapor phase polymerization is also a feasible strategy to obtain highly conductive PEDOT layers [112]. Substrates coated with iron (III)-tosylate as the oxidant were placed in a special reaction chamber, where the evaporation of EDOT and its polymerization on the substrate was performed.

In situ PEDOT refers to the oxidative polymerization of EDOT to PEDOT under film-forming conditions. Up to now, in situ PEDOT represents the highest achievable conductivity of PTh and that is the reason of great practical and commercial value [113]. The overall reaction of EDOT to in situ PEDOT is roughly similar with the oxidative polymerization mechanism for alkylthiophenes [114]. It can be separated into two principal steps: (i) oxidative polymerization of EDOT to the neutral, undoped PTh and (ii) p-type doping of the neutral polymer to the conductive polycations by excess ionic oxidants. Detailed investigations of the reaction kinetics reveal an even more complex reaction path or mechanism [115]. The reaction starts with the EDOT monomers oxidation to the radical cations, which is the slowest and rate determining step. Then the free radical cations rapidly dimerizes. Subsequently, a fast end group oxidation of EDOT oligomers follows, starting with the dimer oxidation. The recombination of two radical cationic end groups results in formation of higher oligomers. Finally, oligomers or polymers are doped by further oxidation to polaron or bipolaron state, achieving the conductive PEDOT. Generally, this oxidative in situ polymerization is preferably carried out with metal ionic oxidants like iron (III), manganese (IV), or similar metal ions in a suitable higher oxidation state. Other metal salt oxidants and peroxides which were mentioned in above oxidative polymerization are also alternatives. Considering the limited solubility of EDOT in water and a suitable reactivity, the well alcohol-soluble iron salts of sulfonic acids are preferred oxidants. Especially, the p-toluenesulfonate has been established as a very suitable anion and the corresponding iron (III) toluenesulfonate has become the most widely used oxidant for EDOT in the preparation of in situ PEDOT layers [116].

2.4.2 Counterions for PEDOT

Counterions are always present in all oxidized forms of PEDOT for charge balancing. It is well known that the counterion is often anion which directly derives from the oxidation agent added in the form of salts to the reaction mixture. PEDOT:counterion complex would have formed in the mixture during the process of chemical polymerization. The choice of counterions is limited only by their solubility and stability under the reaction conditions. Various counterions may also have an effect on the oxidation or polymerization level of the polymer.

Jonas and Heywang originally developed some of the most important oxidation agents, including iron (III) chloride, iron (III) tosylate, or ammonium peroxodisulfate [117]. The corresponding ions of chloride, tosylate, peroxodisulfate, or sulfate serve as counterions. As discussed in the above chemical polymerization of PEDOT, iron (III) salts especially the iron (III)-sulfonates were proven to be particularly useful oxidation agents. One of the most prominent cases is Fe (III)-tosylate due to its well solubility in alcohol solvents and readily available. Mixtures of Fe (III)-tosylate solutions with EDOT were easily prepared and resulted in a well-controlled deposition of conductive films with conductivities as high as 1000 $S\,cm^{-1}$ [111]. Furthermore, the tosylate counterion was proven to promote the chemical polymerization of EDOT. Kirchmeyer and Jonas also reported the use of Fe (III) camphor sulfonate, Fe (III) phenol sulfonate, and Fe (III) methanesulfonate as oxidation agents and the camphor sulfonate can result in highly conductive films with a specific conductivity of more than 1000 $S\,cm^{-1}$ [118]. The use of peroxodisulfate as oxidation agent would ultimately result in films with sulfate as counterion as reported by Jonas and Heywang [117]. However, Kirchmeyer and Reuter later showed that the use of sulfate as the counterion for the polymerization of EDOT was insufficient and additional counterions were always required [115]. Polystyrenesulfonic acid (PSS) is the most prominent polymer counterion for PEDOT. A suitable and well-defined route to an industrially applicable form of PEDOT is the polymerization of EDOT in the presence of PSS as a template polymer. Next section will describe the complex PEDOT:PSS and its application in details. An early work of Jonas and Krafft on PEDOT dispersions showed that PEDOT:PSS could be regarded as a complex with two types of counterions [119]. The complex contained sulfate ions which can be obtained from both the oxidation agents Fe (III) sulfate and potassium peroxodisulfate as well as polystyrenesulfonic acid. Note that the sulfate ion can easily be removed through ion exchange resins [120].

2.4.3 PEDOT:PSS

PEDOT:PSS is the most important and successful conductive polymer in terms of practical application. It possesses many outstanding properities, such as good film-forming ability by versatile fabrication techniques, high transparency in the visible range, high and tunable conductivity, tunable work function, and excellent thermal stability [115, 121]. Strictly speaking, PEDOT:PSS is a polyelectrolyte complex consisting of positively charged conjugated PEDOT and negatively charged saturated PSS. The PSS in the complex not only acts as counterions for charge balancing but also a polymer surfactant aiding for the PEDOT segments dispersed in water or other solvents. PSS intrinsically dissolves in water easily and is commercially available in a large range of molecular weights with different polydispersities. Similar to most conductive polymers, the pristine PEDOT polycation is almost not soluble in any solvents. The polyelectrolyte complex with a loosely crosslinked and highly swollen polymer gel network structure is formed in situ as the polycation chain generates and grows in the presence of PSS. In addition, the solvent of choice for the synthesis of PEDOT:PSS complex is water, which is a poor solvent for the monomer EDOT. The high polarity of PSS can increase the solubility of EDOT in water [122]. Moreover, the acidity of PSS caused by the sulfonic acid groups is beneficial for the EDOT polymerization since acid acts as a catalyst for the reaction [115].

PEDOT:PSS aqueous dispersion was first commercialized under the trade name of Baytron® by Bayer AG, then by H.C. Starck and currently by Heraeus under the trade name of Clevios™. The PEDOT:PSS dispersions are classified into different grades basing on the solid content and the ratio of PEDOT to PSS. These will strongly influence the particles size and distribution, viscosity as well as conductivity and transparency after film formation of the dispersion [115] The important properties of the commercial PEDOT:PSS Clevios dispersions are summarized in Table 2.1.

The commercially available PEDOT:PSS aqueous dispersion is a deep-blue opaque solution. It can be deposited and readily form continuous thin film on either rigid or flexible substrates by all common techniques employed for the deposition of waterborne coatings (such as spin coating, doctor blading, slot die coating, inkjet printing, etc.). The PEDOT:PSS film is smooth and the thickness can also be controlled, making these films attractive as cladding layer for thin film device applications. Generally, PEDOT:PSS thin films are highly transparent in the visible region. The absorption spectrum of PEDOT:PSS is almost identical to that of in situ chemically polymerized PEDOT without PSS. The addition of PSS or other additives like high boiling solvents will not significantly influence the optical properties of PEDOT. While the conductivity of PEDOT:PSS layers exhibits a wide range from 10^{-5} to 10^3 S cm^{-1} and is closely related to the synthetic conditions, processing additives, or posttreatment. The work function (W_f) of PEDOT:PSS films is 5.0~5.2 eV. Dedoping of PEDOT:PSS can lead to a decrease of W_f at various degrees. The tunable work function and conductivity suggest a suitable use of PEDOT:PSS as different functional layers for charge transfer and collection in the electronics. Additional, PEDOT:PSS also possesses good thermal-, photo-, and mechanical-stability.

Table 2.1 The properties of commercial PEDOT:PSS dispersons in water.

Trade name	Solids content in water (wt.%)	PEDOT:PSS ratio (w/w)	Viscosity at 20 °C (mPa)	Particle size d_{50} (nm)	Conductivity ($S cm^{-1}$)
Clevios P	1.3	1:2.5	80	80	<10
Clevios PH	1.3	1:2.5	20	30	<10
Clevios PVP AI 4083	1.5	1:6	10	40	10^{-3}
Clevios PVP CH800	2.8	1:20	15	25	10^{-5}
Clevios PH500	1.1	1:2.5	25	30	500[a]
Clevios PH750	1.1	1:2.5	25	30	750[a]
Clevios PH1000	1.1	1:2.5	30	30	1000[a]

[a] Conductivities of Clevios PH500, PH750, and PH1000 are measured for dispersions containing 5% dimethyl sulfoxide.

All the above-mentioned characteristics render PEDOT:PSS a promising candidate in flexible energy conversion and storage fields.

2.4.4 Applications in Organic Electronics

2.4.4.1 As an Electrode in Organic Solar Cells

Indium tin oxide (ITO) is the current work horse transparent electrode for organic solar cells (OSCs). Despite its good combination of transparency and conductivity, ITO is too expensive and rigid. The solution-processable PEDOT:PSS becomes a main candidate as transparent electrode owing to its unique physical, chemical, and electrical properties.

In 2002, Zhang et al. first demonstrated the sorbitol-added PEDOT:PSS with a conductivity of ~10 S cm^{-1} could be employed as the transparent electrodes to replace ITO for OSCs [123]. Subsequently, Ouyang et al. further improved the conductivity by adding ethylene glycol or m-erythritol into the PEDOT:PSS dispersions [124, 125]. They obtained a better OSCs performance. Since then, different strategies, such as adding high boiling polar solvents (dimethyl sulfone), posttreatment with acid (sulfuric acid, phosphoric acid, or organic acids), or surfactant, have been used for enhancing its conductivity to meet the electrode requirement for practical application [126–129]. Recently, benefiting from the development of the state-of-the-art non-fullerene acceptors, ITO-free OSCs with PEDOT:PSS as bottom transparent electrode have yielded a benchmark power conversion efficiency over 10% [130].

A highly conductive PEDOT:PSS can also be used to replace traditional top metal electrodes. With regard to this case, accurate pattern of the solution-processed PEDOT: PSS films is important. The top opaque metal electrodes are generally patterned via shadow masking during preparation of the OSCs. However, PEDOT:PSS films are typically prepared by spin coating, which makes fine patterning difficult. It was reported that film transfer lamination is an effective technique to pattern the PEDOT:PSS electrode [131]. A schematic diagram is shown in Figure 2.4a. The mechanism is mainly to utilize the different adhesion between PEDOT:PSS film to the transfer medium (polydimethylsiloxane) and the target surface (organic active layer). Before transfer printing, the PEDOT:PSS film is cut into the desired shape and size. Note that the hydrophilic PEDOT:PSS is challenging to deposit on the top of the hydrophobic organic layers. To overcome this wettability issue, several approaches have been utilized including inserting an amphiphilic interlayer, adding surfactant into PEDOT:PSS, and plasma treatment of the active layer surfaces. Owing to the fact that PEDOT:PSS electrodes are solution-processable, vacuum-free, optically transparent, possess tunable work function, and conductivity, and can be patterned with

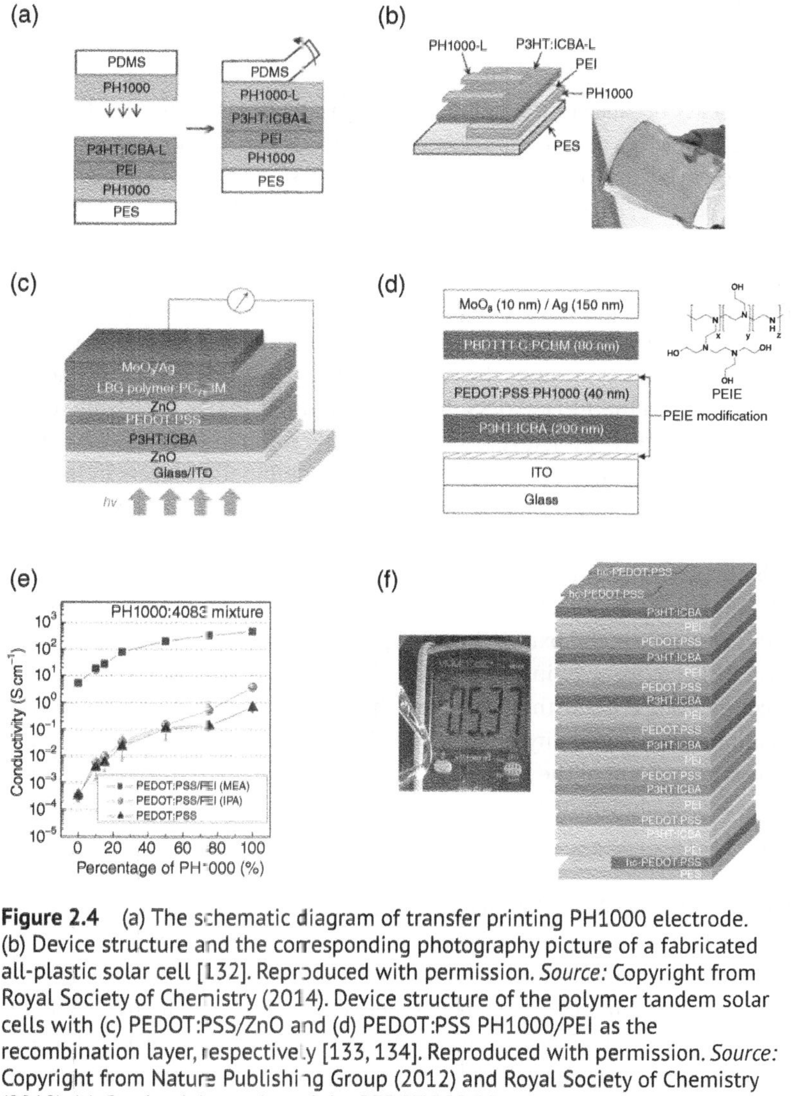

Figure 2.4 (a) The schematic diagram of transfer printing PH1000 electrode. (b) Device structure and the corresponding photography picture of a fabricated all-plastic solar cell [132]. Reproduced with permission. *Source:* Copyright from Royal Society of Chemistry (2014). Device structure of the polymer tandem solar cells with (c) PEDOT:PSS/ZnO and (d) PEDOT:PSS PH1000/PEI as the recombination layer, respectively [133, 134]. Reproduced with permission. *Source:* Copyright from Nature Publishing Group (2012) and Royal Society of Chemistry (2012). (e) Conductivity tuning of the PEDOT:PSS/PEI interconnecting layer. (f) All-solution-processed all-plastic multijunction cells displayed a V_{OC} of 5.37V [135]. *Source:* Reproduced with permission. Copyright from Royal Society of Chemistry (2016).

the transfer printing technique, both ITO electrode and metal electrode can be replaced by PEDOT:PSS films. Zhou et al. first reported an all-plastic OSCs with an architecture: plastic substrates/PEDOT:PSS/PEI/active layer/PEDOT:PSS, as shown in Figure 2.4b [132].

2.4.4.2 Buffer Layer in Organic Solar Cells

PEDOT:PSS (commonly Clevios PVP AI 4083) is the dominant solution-processable hole transport layer (HTL) for anode modification in conventional OSCs because of the following reasons: (i) It has high work function of 5. eV and has been demonstrated as very efficient hole extractor from most active layers; (ii) The films are easily processed from aqueous solutions by spin coating; (iii) PEDOT:PSS dispersion is commercially available; (iv) The films have good air stability and reproducibility; (v) The films are highly transparent and do not affect the transmittance of ITO [136, 137]. However, it is difficult to deposit the PEDOT:PSS aqueous dispersion onto the hydrophobic active layers in an inverted architecture. Current feasible deposition strategy is mainly through the addition of a small amount of surfactant (<1 wt.%) into the PEDOT:PSS dispersion and plasma or UV-Ozone treatment of the active layer surfaces to tune its hydrophilic [138, 139]. Another application is used to construct the charge recombination layers in tandem OSCs. The charge recombination layers, also referred to interconnecting layers, function as the charge recombination zone to connect the two subcells. It generally consists of a p-type material for hole transport and an n-type material for electron transport. The PEDOT:PSS is often combined with the n-type metal oxide (ZnO, TiO_2) or the low work function modifier (PEI, PEIE) to form an interconnecting layer (Figure 2.4c and d) [133, 134]. Recently, Tong et al. further optimized the conductivity of the charge recombination layers via modifying the PEDOT:PSS formulation and selecting the processing solvent of the PEI to fabricate a multi-junction (up to 7-junction, 22 layers, as shown in Figure 2.4e and f) all-plastic solar cells with high yields [135].

2.4.4.3 Polymer-Based Organic Thermoelectric Generators

PEDOT:PSS has attracted considerable attentions for the thermoelectric (TE) application due to its environmental stability, aqueous solution processability, high electrical conductivity, and extremely low thermal conductivity. The TE characteristics of PEDOT:PSS at various forms (including pellets [140], 1-D nanostructure [141], freestanding films [142], and their composites [143]) have been extensively investigated and made some outstanding progresses. The figure of merit, (ZT) of pure PEDOT:PSS is only 0.0017, limited by the intrinsically low Seebeck coefficient of the conjugated polymers [144]. To enhance the ZT of PEDOT, many methods are employed for improving the Seebeck coefficient and electrical conductivity of PEDOT thin films. Ouyang and co-workers reported the conductivity enhancement of PEDOT:PSS through addition of high-boiling solvents (such as dimethyl formamide, dimethyl sulfoxide, and polyalcohols) or a posttreatment with acids. They enhanced the TE property of PEDOT:PSS films through sequential posttreatments with common acids and bases. The optimal films exhibited a Seebeck coefficient up to 39.2 μV K^{-1}, electrical conductivity of 2170 S cm^{-1}, and a corresponding PF of 334 μW (m^{-1} K^{-2}) (Figure 2.5a and b) [145]. A higher electrical conductivity over 4000 S cm^{-1} with an estimated ZT of 0.08 was also achieved by sulfuric acid posttreatment on the films [147]. The breakthrough was then achieved by Crispin and co-workers [148]. They replaced PSS anion with tosylate anion, resulting in an enhanced electrical conductivity over 1000 S cm^{-1} due to the reduction of insulating polyanion phases. Furthermore, they de-doped the highly conductive PEDOT:tosylate with tetrakis(dimethylamino)-ethylene to optimize the electrical conductivity and Seebeck coefficient. A record-high power factor of 324 μW m^{-1} K^{-2} is achieved from PEDOT-Tos materials at the oxidation level of 22% due to its very high Seebeck coefficient of 780 μV K^{-1}. The corresponding ZT of 0.25 was the highest ever reported in polymer thermoelectric materials at that time. Very recently, the ZT is further increased to 0.42 for PEDOT:PSS film at room temperature by adding DMSO and EG to the commercial PEDOT:PSS dispersion [149]. Inspired by the achievements, we employed a highly conductive structure-ordered PEDOT:PSS film to fabricate a free-standing polymeric TE device [142]. Under a small temperature gradient of 29 K, the TE device demonstrated a maximum output power density of 99 ± 18.7 μW cm^{-2} (Figure 2.5c). In addition, a fivefold output power was demonstrated by series connecting five devices into a flexible thermoelectric module (Figure 2.5d). Dong et al. successfully prepared a flexible module composed of PEDOT:PSS as the p-type leg while PEDOT:PSS coated on top of the ITO as n-type leg, on a PET substrate [146]. As presented in Figure 2.5e, the thermoelectric modules can function as a micro thermoelectric generator to produce thermovoltage via the temperature gradience produced by a human hand or warm water.

2.5 Processing Technics for Flexible Thermoelectric Generators

Compared with the current dominant commercial Bi_2Te_3-based thermoelectric applications, one of the most attractive advantage of organic thermoelectrics is their easy processing, which can be processed from solutions at low temperature or room temperature. Solution-processability is a key to scalable and cost-effective fabrication of the thermoelectric generators. Flexible thermoelectric generators can be realized by depositing the organic materials on the flexible substrates such as polyethylene terephthalate (PET), polyethylene naphthalate (PEN), and polyimide (PI). In addition,

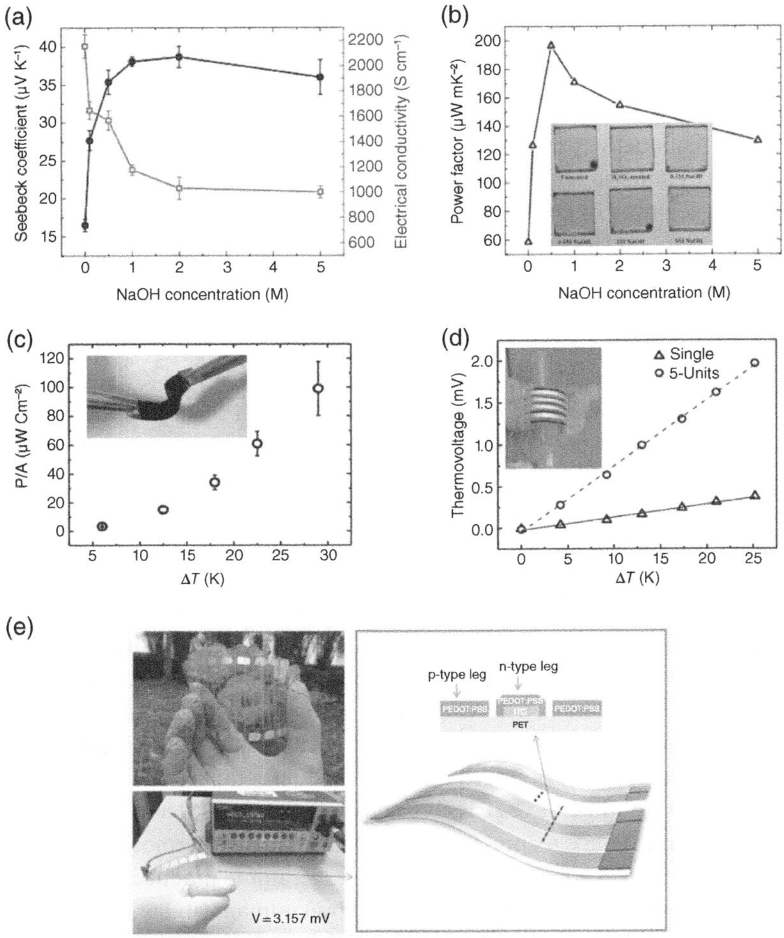

Figure 2.5 (a) Seebeck coefficient, electrical conductivity, and (b) power factor of the H_2SO_4-treated PEDOT:PSS films at various NaOH concentration treatment conditions [145]. *Source:* Reproduced with permission. Copyright from Wiley-VCH (2017). (c) Power output per cross-sectional area of the free-standing TE device-based highly conductive structure-ordered PEDOT:PSS films and (d) the voltage-temperature gradient characteristics of the corresponding single component and module [142]. *Source:* Reproduced with permission. Copyright from Wiley-VCH (2018). (e) The flexible and transparent organic–inorganic hybrid thermoelectric module on the PET substrate with PEDOT:PSS as the p-type legs and the ITO-PEDOT:PSS as the n-type legs [146]. *Source:* Reproduced with permission. Copyright from American Chemical Society (2018).

organic materials have better biocompatibility than inorganic materials, making them the best candidates for flexible and wearable thermoelectric devices for applications related to human body and biosensing.

The techniques applied to fabricate flexible thermoelectric devices in the literature include spin coating, doctor blading, screen printing, and inject printing, etc. [150]. Spin coating is the most commonly used method in the lab for flexible thermoelectric generators. In the operations, a liquid is dropped onto a substrate followed by a substrate, spinning at a selected speed. It gives excellent reproducible and uniform films. Film thicknesses are easily tunable by changing the spinning speed and solution concentration. Though the spin coating is widely used for laboratory-size (about 10 mm^2) devices, there are several drawbacks: (i) This technique requires substrates to be handled individually and is not compatible with the roll-to-roll processing; (ii) It does not give required patterning in thermoelectric modules; (iii) The technique loses the solution/ink as high as 90% during the processing. Therefore, the spin coating technique does not meet the requirement of the low-cost large-area and high-throughput production of the thermoelectric generators. Doctor blading is another coating technique that has been demonstrated in the literature for the fabrication of thermoelectric generators (Figure 2.6a). The doctor blading technique is to place a sharp blade at a fixed distance from the substrate surface that is to be coated. The coating solution is then placed in front of the blade. The blade was moved linearly across the substrate leaving a thin wet film. The film thickness depends on the concentration, viscosity of the solution and the surface energy of the substrates. The doctor blading technique allows for a well-defined thickness and almost no loss of ink during the film formation. It can also be integrated with the roll-to-roll processing. In spite of these advantages, the doctor blading strategy is not used as widely as spin coating method. That is mainly because practically it takes time and materials for getting the right condition for film processing. Furthermore, the films prepared by doctor blading dries much slower than those prepared by spin coating. The slow dry of the ink could significantly influence the film morphology and tend to form aggregates and crystalline structure that will affect the device performance and reproducibility. Screen printing is also a commonly used method for its straightforward process (Figure 2.6b). It is a printing technique whereby a pre-patterned mesh is used to transfer ink onto a substrate. A blade or squeegee is moved across the screen to fill the open mesh apertures with ink, and a reverse stroke then causes the screen to touch the substrate momentarily along a line of contact. This causes the ink to wet the substrate and be pulled out of the mesh apertures as the screen springs back after the blade has passed. The screen printing

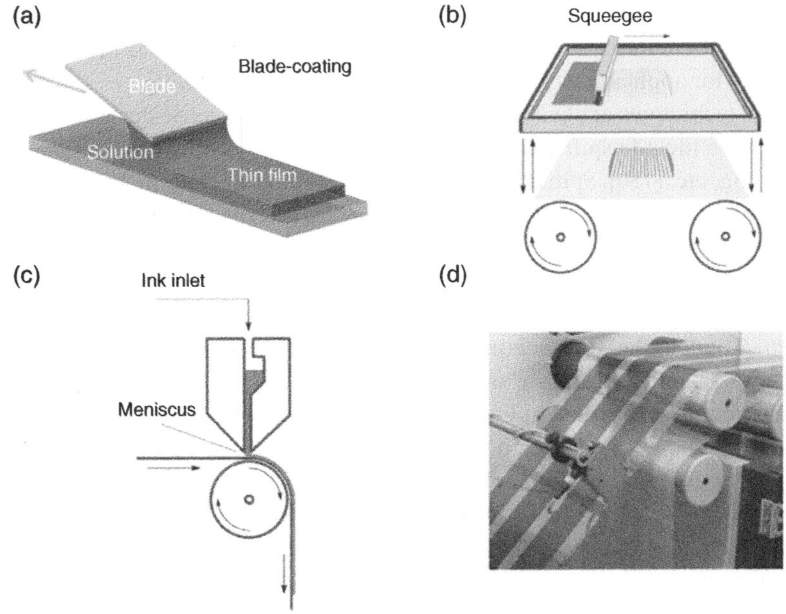

Figure 2.6 Schematic of (a) doctor blade coating, (b) screen printing, and (c) inkjet printing. (d) A photograph of fabricated flexible films [151]. *Source:* Reproduced with permission. Copyright from Elsevier (2012).

technique has long history developed about a thousand years ago in China. It has been introduced for the fabrication of conductive layers for high performance thermoelectric modules in early 2000s [152]. The potential of this technique can be magnified when integrated into the roll-to-roll (R2R) production line that is capable of continuously manufacturing meter-long modules. Søndergaard et al. developed an automatic R2R process and screen printed as many as 7000 thermoelectric elements per square meter on PET foil [153]. Each pair of elements composed of a screen-printed p-type PEDOT:PSS leg and a second flexo-printed leg-based Ag ink. Inkjet printing, driven forward by the typically used office inkjet printer, has also been used to prepare the conductive layer for flexible thermoelectric generators (Figure 2.6c and d). The original ink cartridge is replaced by the thermoelectric material formulation which is critical part for the inkjet processing. It is typically by mechanical compression of the ink from a nozzle. Then the droplet is electrostatically charged. Surface tension of the ink has to be sufficiently high to generate a stream of the droplet. Typically, different solvents and additives are added to the solution to get the right formulations for high-quality printed films. The inkjet printing process has the advantages of less material waste, high-resolution patterning, and convenience to create complex patterns.

2.6 Conclusions and Perspectives

Over the last couple of years, organic thermoelectrics was under tremendous progress. Conjugated polymers have shown significant potential for thermoelectric application owing to the advantages of easier processability and intrinsically low thermal conductivity as well as flexibility, light weight, and material abundance. Recently, the maximum ZT of p-type polymers have been obtained 0.42 at room temperature and the best one reached 0.2 at 167 °C for n-type organic materials. Large-area organic thermoelectric modules on flexible substrates have also been demonstrated via roll-to-roll processing. The rapid progress in this field is mostly based on the progress in the materials science and production technologies. However, current accomplishment is not enough to compete with their inorganic counterparts. The efficiency of polymeric thermoelectric materials is mainly restricted by the relatively low power factor in comparison to those of inorganics. A further improvement is needed to realize a high ZT (over 0.4) on flexible substrates in large area. One key step is to understand the complex structure–property relationship and develop more versatile and efficient materials. Second is to optimize the doping level for a balance of the electrical conductivity and Seebeck coefficient for a maximum power factor. Final step is to select rational coating or manufacturing techniques to gain improved control over the solid-state morphology. Additional, long-term stability and sustainability of the flexible devices in order to enable the transition to a truly viable technology should also be systematically studied. The organic thermoelectric devices may have huge potential applications for flexible and wearable energy harvesters, especially as low-power units in the future.

Acknowledgments

This work was jointly supported by the National Natural Science Foundation of China (Grant Nos. 51773072 and 21474035), the Fundamental Research Funds for the Jiaxing University (Grant No. CD70519048), and the Jiaxing Public Welfare Research Program in 2019 (No. 2019AY11007).

References

1 McGrail, B.T., Sehirlioglu, A., and Pentzer, E. (2015). Polymer composites for thermoelectric applications. *Angew. Chem. Int. Ed. 54*: 1710–1723.

2 Wang, L., Liu, Y., Zhang, Z. et al. (2017). Polymer composites-based thermoelectric materials and devices. *Composites Part B 122*: 145–155.

3 He, M., Qiu, F., and Lin, Z.Q. (2013). Towards high-performance polymer-based thermoelectric materials. *Energy Environ. Sci. 6*: 1352–1361.

4 Zaumseil, J. and Sirringhaus, H. (2007). Electron and ambipolar transport in organic field-effect transistors. *Chem. Rev. 107*: 1296–1323.

5 Burroughes, J.H., Bradley, D.D., Brown, A. et al. (1990). Light-emitting diodes based on conjugated polymers. *Nature 347*: 539.

6 Günes, S., Neugebauer, H., and Sariciftci, N.S. (2007). Conjugated polymer-based organic solar cells. *Chem. Rev. 107*: 1324–1338.

7 Bubnova, O. and Crispin, X. (2012). Towards polymer-based organic thermoelectric generators. *Energy Environ. Sci. 5*: 9345.

8 Yang, J., Yip, H.-L., and Jen, A.K.Y. (2013). Rational design of advanced thermoelectric materials. *Adv. Energy Mater. 3*: 549–565.

9 Zhang, Q., Sun, Y., Xu, W., and Zhu, D. (2014). Organic thermoelectric materials: emerging green energy materials converting heat to electricity directly and efficiently. *Adv. Mater. 26*: 6829–6851.

10 Chiang, C.K., Fincher, C.R., Park, Y.W. et al. (1977). Electrical conductivity in doped polyacetylene. *Phys. Rev. Lett. 39*: 1098–1101.

11 Shirakawa, H., Louis, E.J., MacDiarmid, A.G. et al. (1977). Synthesis of electrically conducting organic polymers: halogen derivatives of polyacetylene,(CH)x. *J. Chem. Soc. Chem. Commun. 16*: 578–580.

12 Kim, J., Lee, J., You, J. et al. (2016). Conductive polymers for next-generation energy storage systems: recent progress and new functions. *Mater. Horiz. 3*: 517–535.

13 Letheby, H. (1862). XXIX.-On the production of a blue substance by the electrolysis of sulphate of aniline. *J. Chem. Soc. 15*: 161–163.

14 Akamatu, H., Inokuchi, H., and Matsunaga, Y. (1954). Electrical conductivity of the perylene-bromine complex. *Nature 173*: 168–169.

15 Ferraris, J., Cowan, D., Walatka, V.t., and Perlstein, J. (1973). Electron transfer in a new highly conducting donor-acceptor complex. *J. Am. Chem. Soc. 95*: 948–949.

16 Bolto, B.A., McNeill, R., and Weiss, D. (1963). Electronic conduction in polymers. III. electronic properties of polypyrrole. *Aust. J. Chem. 16*: 1090–1103.

17 De Surville, R., Jozefowicz, M., Yu, L. et al. (1968). Electrochemical chains using protolytic organic semiconductors. *Electrochim. Acta 13*: 1451–1458.

18 Diaz, A.F. and Logan, J.A. (1980). Electroactive polyaniline films. *J. Electroanal. Chem. Interfacial Electrochem.* *111*: 111–114.

19 Naarmann, H. (1982). Synthese elektrisch leitfähiger polymerer. *Die Angewandte Makromolekulare Chemie 109*: 295–338.

20 BOŽOVIĆ, I. (1987). Polyacetylene: helices, heavy solitons and hyperconductivity. *Mod. Phys. Lett. B 01*: 81–88.

21 Saxman, A., Liepins, R., and Aldissi, M. (1985). Polyacetylene: Its synthesis, doping and structure. *Prog. Polym. Sci. 11*: 57–89.

22 Feast, W., Tsibouklis, J., Pouwer, K. et al. (1996). Synthesis, processing and material properties of conjugated polymers. *Polymer 37*: 5017–5047.

23 Shirakawa, H. and Ikeda, S. (1971). Infrared spectra of poly (acetylene). *Polym. J. 2*: 231.

24 Rasmussen, S.C. (2018). Doped polyacetylene. In: *Acetylene and Its Polymers*, 109–124. Springer.

25 Mortensen, K., Thewalt, M., Tomkiewicz, Y. et al. (1980). Non-ohmic behavior of cis-polyacetylene doped with AsF_5. *Phys. Rev. Lett. 45*: 490.

26 Bhadra, S., Khastgir, D., Singha, N.K., and Lee, J.H. (2009). Progress in preparation, processing and applications of polyaniline. *Prog. Polym. Sci. 34*: 783–810.

27 Kang, E., Neoh, K., and Tan, K. (1998). Polyaniline: interesting a polymer with many intrinsic redox states. *Prog. Polym. Sci. 23*: 211–324.

28 MacDiarmid, A., Chiang, J., Richter, A., and Epstein, A.J. (1987). Polyaniline: a new concept in conducting polymers. *Synth. Met. 18*: 285–290.

29 MacDiarmid, A.G. (2001). "Synthetic metals": a novel role for organic polymers. *Curr. Appl. Phys. 1*: 269–279.

30 Huang, L.-M., Chen, C.-H., and Wen, T.-C. (2006). Development and characterization of flexible electrochromic devices based on polyaniline and poly(3,4-ethylenedioxythiophene)-poly(styrene sulfonic acid). *Electrochim. Acta 51*: 5858–5863.

31 Virji, S., Huang, J., Kaner, R.B., and Weiller, B.H. (2004). Polyaniline nanofiber gas sensors: examination of response mechanisms. *Nano Lett. 4*: 491–496.

32 Wessling, B. (2010). New insight into organic metal polyaniline morphology and structure. *Polymers 2*: 786–798.

33 Bleda-Martínez, M.J., Morallón, E., and Cazorla-Amorós, D. (2007). Polyaniline/porous carbon electrodes by chemical polymerisation: effect of carbon surface chemistry. *Electrochim. Acta 52*: 4962–4968.

34 Wang, L.-X., Li, X.-G., and Yang, Y.-L. (2001). Preparation, properties and applications of polypyrroles. *React. Funct. Polym. 47*: 125–139.

35 Rühe, J., Ezquerra, T., and Wegner, G. (1989). New conducting polymers from 3-alkylpyrroles. *Synth. Met. 28*: 177–181.

36 Lee, J.-W., Serna, F., Nickels, J., and Schmidt, C.E. (2006). Carboxylic acid-functionalized conductive polypyrrole as a bioactive platform for cell adhesion. *Biomacromolecules 7*: 1692–1695.

37 Stanke, D., Hallensleben, M.L., and Toppare, L. (1995). Oxidative polymerization of some N-alkylpyrroles with ferric chloride. *Synth. Met. 73*: 267–272.

38 Naarmann, H. (1987). *New aspects on intrinsically conducting organic systems and their synthesis*, Makromolekulare Chemie. *Macromol. Symp.* Wiley Online Library; pp 1–15.

39 Huang, Z.-B., Yin, G.-F., Liao, X.-M., and Gu, J.-W. (2014). Conducting polypyrrole in tissue engineering applications. *Front. Mater. Sci. 8*: 39–45.

40 Kakuda, S., Momma, T., Osaka, T. et al. (1995). Ambient-temperature, rechargeable, all-solid lithium/polypyrrole polymer battery. *J. Electrochem. Soc. 142*: L1–L2.

41 Freund, M.S. and Lewis, N.S. (1995). A chemically diverse conducting polymer-based "electronic nose". *Proc. Natl. Acad. Sci. 92*: 2652–2656.

42 Culebras, M., Uriol, B., Gómez, C.M., and Cantarero, A. (2015). Controlling the thermoelectric properties of polymers: application to PEDOT and polypyrrole. *Phys. Chem. Chem. Phys. 17*: 15140–15145.

43 Tourillon, G. and Garnier, F. (1983). Stability of conducting polythiophene and derivatives. *J. Electrochem. Soc. 130*: 2042–2044.

44 Bartuš, J. (1991). Electrically conducting thiophene polymers. *J. Macromol. Sci. Part A: Chem. 28*: 917–924.

45 Loponen, M.T., Taka, T., Laakso, J. et al. (1991). Doping and dedoping processes in poly (3-alkylthiophenes). *Synth. Met. 41*: 479–484.

46 Qiao, X., Wang, X., and Mo, Z. (2001). The $FeCl_3$-doped poly (3-alkyithiophenes) in solid state. *Synth. Met. 122*: 449–454.

47 Osaka, I. and McCullough, R.D. (2008). Advances in molecular design and synthesis of regioregular polythiophenes. *Acc. Chem. Res. 41*: 1202–1214.

48 Granström, M., Berggren, M., Pede, D. et al. (1997). Self organizing polymer films-a route to novel electronic devices based on conjugated polymers. *Supramol. Sci. 4*: 27–34.

49 McQuade, D.T., Pullen, A.E., and Swager, T.M. (2000). Conjugated polymer-based chemical sensors. *Chem. Rev. 100*: 2537–2574.

50 Mehmood, U., Al-Ahmed, A., and Hussein, I.A. (2016). Review on recent advances in polythiophene based photovoltaic devices. *Renew. Sust. Energ. Rev. 57*: 550–561.

51 Nielsen, C.B. and McCulloch, I. (2013). Recent advances in transistor performance of polythiophenes. *Prog. Polym. Sci. 38*: 2053–2069.

52 Bharti, M., Singh, A., Samanta, S., and Aswal, D.K. (2018). Conductive polymers for thermoelectric power generation. *Prog. Mater. Sci. 93*: 270–310.

53 Le, T.H., Kim, Y., and Yoon, H. (2017). Electrical and electrochemical properties of conducting polymers. *Polymers (Basel) 9*.

54 Bredas, J.L. and Street, G.B. (1985). Polarons, bipolarons, and solitons in conducting polymers. *Acc. Chem. Res. 18*: 309–315.

55 Chen, Y., Zhao, Y., and Liang, Z. (2015). Solution processed organic thermoelectrics: towards flexible thermoelectric modules. *Energy Environ. Sci. 8*: 401–422.

56 Roncali, J. (2007). Molecular engineering of the band gap of π-conjugated systems: facing technological applications. *Macromol. Rapid Commun. 28*: 1761–1775.

57 Brédas, J.L. (1985). Relationship between band gap and bond length alternation in organic conjugated polymers. *J. Chem. Phys. 82*: 3808–3811.

58 Bradley, D.D.C., Evans, G.P., and Friend, R.H. (1987). Characterisation of poly(phenylenevinylene) by infrared and optical absorption. *Synth. Met. 17*: 651–656.

59 Wudl, F., Kobayashi, M., and Heeger, A. (1984). Poly (isothianaphthene). *J. Organomet. Chem. 49*: 3382–3384.

60 Kitamura, C., Tanaka, S., and Yamashita, Y. (1996). Design of narrow-bandgap polymers. Syntheses and properties of monomers and polymers containing aromatic-donor and o-quinoid-acceptor units. *Chem. Mater. 8*: 570–578.

61 Brédas, J.-L., Norton, J.E., Cornil, J., and Coropceanu, V. (2009). Molecular understanding of organic solar cells: the challenges. *Acc. Chem. Res. 42*: 1691–1699.

62 Zhang, Q.T. and Tour, J.M. (1998). Alternating donor/acceptor repeat units in polythiophenes. Intramolecular charge transfer for reducing band gaps in fully substituted conjugated polymers. *J. Am. Chem. Soc. 120*: 5355–5362.

63 Noriega, R., Rivnay, J., Vandewal, K. et al. (2013). A general relationship between disorder, aggregation and charge transport in conjugated polymers. *Nat. Mater. 12*: 1038–1044.

64 Hou, J., Tan, Z.a., Yan, Y. et al. (2006). Synthesis and photovoltaic properties of two-dimensional conjugated polythiophenes with bi (thienylenevinylene) side chains. *J. Am. Chem. Soc. 128*: 4911–4916.

65 Ye, L., Zhang, S., Huo, L. et al. (2014). Molecular design toward highly efficient photovoltaic polymers based on two-dimensional conjugated benzodithiophene. *Acc. Chem. Res. 47*: 1595–1603.

66 Zhan, C. and Yao, J. (2016). More than conformational "twisting" or "coplanarity": molecular strategies for designing high-efficiency nonfullerene organic solar cells. *Chem. Mater. 28*: 1948–1964.

67 Chen, H., Guo, Y., Yu, G. et al. (2012). Highly pi-extended copolymers with diketopyrrolopyrrole moieties for high-performance field-effect transistors. *Adv. Mater. 24*: 4618–4622.

68 McCulloch, I., Ashraf, R.S., Biniek, L. et al. (2012). Design of semiconducting indacenodithiophene polymers for high performance transistors and solar cells. *Acc. Chem. Res. 45*: 714–722.

69 Tsao, H.N., Cho, D.M., Park, I. et al. (2011). Ultrahigh mobility in polymer field-effect transistors by design. *J. Am. Chem. Soc. 133*: 2605–2612.

70 Facchetti, A. (2011). π-Conjugated polymers for organic electronics and photovoltaic cell applications. *Chem. Mater. 23*: 733–758.

71 Zhang, W., Smith, J., Watkins, S.E. et al. (2010). Indacenodithiophene semiconducting polymers for high-performance, air-stable transistors. *J. Am. Chem. Soc. 132*: 11437–11439.

72 Elschner, A., Kirchmeyer, S., Lovenich, W. et al. (2010). *PEDOT: Principles and Applications of an Intrinsically Conductive Polymer*. CRC Press.

73 McCullough, R.D. (1998). The chemistry of conducting polythiophenes. *Adv. Mater. 10*: 93–116.

74 Cheng, Y.J., Yang, S.H., and Hsu, C.S. (2009). Synthesis of conjugated polymers for organic solar cell applications. *Chem. Rev. 109*: 5868–5923.

75 Mei, J. and Bao, Z. (2013). Side chain engineering in solution-processable conjugated polymers. *Chem. Mater. 26*: 604–615.

76 Lussem, B., Keum, C.M., Kasemann, D. et al. (2016). Doped organic transistors. *Chem. Rev. 116*: 13714–13751.

77 Sariciftci, N.S., Heeger, A., and Nalwa, H. (1997). *Handbook of Organic Conductive Molecules and Polymers*, vol. *1*, 414. *John Wiley and Sons Ltd.*

78 Tietze, M.L., Rose, B.D., Schwarze, M. et al. (2016). Passivation of molecular n-doping: exploring the limits of air stability. *Adv. Funct. Mater. 26*: 3730–3737.

79 Parthasarathy, G., Shen, C., Kahn, A., and Forrest, S.R. (2001). Lithium doping of semiconducting organic charge transport materials. *J. Appl. Phys. 89*: 4986–4992.

80 Fröbel, M., Schwab, T., Kliem, M. et al. (2015). Get it white: color-tunable AC/DC OLEDs. *Light: Sci. Appl. 4*: e247–e247.

81 Benamara, A.A., Galtier, M., and Montaner, A. (1991). N doping of polyacetylene. *Synth. Met. 41*: 45–48.

82 Huang, F., Shih, P.-I., Liu, M.S. et al. (2008). Lithium salt doped conjugated polymers as electron transporting materials for highly efficient blue polymer light-emitting diodes. *Appl. Phys. Lett. 93*: 243302.

83 Nollau, A., Pfeiffer, M., Fritz, T., and Leo, K. (2000). Controlled n-type doping of a molecular organic semiconductor: Naphthalenetetracarboxylic dianhydride (NTCDA) doped with bis(ethylenedithio)-tetrathiafulvalene (BEDT-TTF). *J. Appl. Phys. 87*: 4340–4343.

84 Cotton, F.A., Gruhn, N.E., Gu, J. et al. (2002). Closed-shell molecules that ionize more readily than cesium. *Science 298*: 1971–1974.

85 Tanaka, S., Kanai, K., Kawabe, E. et al. (2005). Doping effect of tetrathianaphthacene molecule in organic semiconductors on their interfacial electronic structures studied by UV photoemission spectroscopy. *Jpn. J. Appl. Phys. 44*: 3760–3763.

86 Vaid, T.P., Lytton-Jean, A.K., and Barnes, B.C. (2003). Investigations of the 9, 10-diphenylacridyl radical as an isostructural dopant for the molecular semiconductor 9, 10-diphenylanthracene. *Chem. Mater. 15*: 4292–4299.

87 Cho, N., Yip, H.-L., Davies, J.A. et al. (2011). In-situ crosslinking and n-doping of semiconducting polymers and their application as efficient electron-transporting materials in inverted polymer solar cells. *Adv. Energy Mater. 1*: 1148–1153.

88 Wei, P., Oh, J.H., Dong, G., and Bao, Z. (2010). Use of a 1 H-benzoimidazole derivative as an n-type dopant and to enable air-stable solution-processed n-channel organic thin-film transistors. *J. Am. Chem. Soc. 132*: 8852–8853.

89 Limmert, M., Zeika, O., Ammann, M. et al. (2012). Heterocyclic radical or diradical, the dimers, oligomers, polymers, dispiro compounds and polycycles thereof, the use thereof, organic semiconductive material and electronic or optoelectronic component. *Google Patents.*

90 Guo, S., Kim, S.B., Mohapatra, S.K. et al. (2012). n-Doping of organic electronic materials using air-stable organometallics. *Adv. Mater. 24*: 699–703.

91 Naab, B.D., Zhang, S., Vandewal, K. et al. (2014). Effective solution- and vacuum-processed n-doping by dimers of benzimidazoline radicals. *Adv. Mater. 26*: 4268–4272.

92 Hu, L., Liu, T., Duan, J. et al. (2017). An amidine-type n-dopant for solution-processed field-effect transistors and perovskite solar cells. *Adv. Funct. Mater. 27*: 1703254.

93 Fabiano, S., Braun, S., Liu, X. et al. (2014). Poly(ethylene imine) impurities induce n-doping reaction in organic (semi)conductors. *Adv. Mater. 26*: 6000–6006.

94 Zhao, X., Madan, D., Cheng, Y. et al. (2017). High conductivity and electron-transfer validation in an n-type fluoride-anion-doped polymer for thermoelectrics in air. *Adv. Mater. 29*: 1606928.

95 Wu, Z., Sun, C., Dong, S. et al. (2016). n-Type water/alcohol-soluble naphthalene diimide-based conjugated polymers for high-performance polymer solar cells. *J. Am. Chem. Soc. 138*: 2004–2013.

96 Patil, A., Heeger, A., and Wudl, F. (1988). Optical properties of conducting polymers. *Chem. Rev. 88*: 183–200.

97 Kröger, M., Hamwi, S., Meyer, J. et al. (2009). P-type doping of organic wide band gap materials by transition metal oxides: a case-study on molybdenum trioxide. *Org. Electron. 10*: 932–938.

98 Chan, C.Y.H., Chow, C.M., and So, S.K. (2011). Using transistor technique to study the effects of transition metal oxide dopants on organic charge transporters. *Org. Electron. 12*: 1454–1458.

99 Meyer, J., Hamwi, S., Schmale, S. et al. (2009). A strategy towards p-type doping of organic materials with HOMO levels beyond 6 eV using tungsten oxide. *J. Mater. Chem. 19*: 702.

100 Lee, J.-H., Kim, H.-M., Kim, K.-B., and Kim, J.-J. (2011). Origin of charge generation efficiency of metal oxide p-dopants in organic semiconductors. *Org. Electron. 12*: 950–954.

101 Cao, Y., Smith, P., and Heeger, A.J. (1992). Counter-ion induced processibility of conducting polyaniline and of conducting polyblends of polyaniline in bulk polymers. *Synth. Met. 48*: 91–97.

102 Duong, D.T., Wang, C., Antono, E. et al. (2013). The chemical and structural origin of efficient p-type doping in P3HT. *Org. Electron. 14*: 1330–1336.

103 Sun, K., Zhang, S., Li, P. et al. (2015). Review on application of PEDOTs and PEDOT:PSS in energy conversion and storage devices. *J. Mater. Sci. Mater. Electron. 26*: 4438–4462.

104 Heywang, G. and Jonas, F. (1992). Poly (alkylenedioxythiophene) s-new, very stable conducting polymers. *Adv. Mater. 4*: 116–118.

105 Jiang, C., Chen, G., and Wang, X. (2012). High-conversion synthesis of poly(3,4-ethylenedioxythiophene) by chemical oxidative polymerization. *Synth. Met. 162*: 1968–1971.

106 Corradi, R. and Armes, S. (1997). Chemical synthesis of poly (3, 4-ethylenedioxythiophene). *Synth. Met. 84*: 453–454.

107 Hupe, J., Wolf, G., and Jonas, F. (1995). DMS-E--a recognised principle with a novel basis. Through-hole contacting of printed circuit boards using conductive polymers. *Galvanotechnik(Germany) 86*: 3404–3411.

108 Im, S.G., Kusters, D., Choi, W. et al. (2008). Conformal coverage of poly (3, 4-ethylenedioxythiophene) films with tunable nanoporosity via oxidative chemical vapor deposition. *ACS Nano 2*: 1959–1967.

109 Jonas, F., Heywang, G., Schmidtberg, W. et al. (1990). Polythiophenes, process for their preparation and their use. *Google Patents*.

110 De Leeuw, D.M., Kraakman, P., Bongaerts, P. et al. (1994). Electroplating of conductive polymers for the metallization of insulators. *Synth. Met. 66*: 263–273.

111 Winther-Jensen, B., Breiby, D.W., and West, K. (2005). Base inhibited oxidative polymerization of 3,4-ethylenedioxythiophene with iron(III) tosylate. *Synth. Met. 152*: 1–4.

112 Levermore, P.A., Chen, L., Wang, X. et al. (2007). Fabrication of highly conductive poly(3,4-ethylenedioxythiophene) films by vapor phase polymerization and their application in efficient organic light-emitting diodes. *Adv. Mater. 19*: 2379–2385.

113 Cho, B., Park, K.S., Baek, J. et al. (2014). Single-crystal poly(3,4-ethylenedioxythiophene) nanowires with ultrahigh conductivity. *Nano Lett. 14*: 3321–3327.

114 Taliani, C., Gebauer, W., and Fichou, D. (1999). *Handbook of Oligo and Polythiophenes*. Weinheim: Wiley-VCH.

115 Kirchmeyer, S. and Reuter, K. (2005). Scientific importance, properties and growing applications of poly(3,4-ethylenedioxythiophene). *J. Mater. Chem. 15*: 2077.

116 Hong, K.H., Oh, K.W., and Kang, T.J. (2005). Preparation and properties of electrically conducting textiles byin situ polymerization of poly(3,4-ethylenedioxythiophene). *J. Appl. Polym. Sci. 97*: 1326–1332.

117 Jonas, F. and Heywang, G. (1988). DE 3813589 (Bayer AG). *Prior: April 22.*

118 Jonas, F. and Kirchmeyer, S. (2006). Method of forming polythiophene dispersions. *Google Patents.*

119 Jonas, F. and Krafft, W. (1994). Polythiophene dispersions, their production and their use. *Google Patents.*

120 Jonas, F., Lerch, K., and Guntermann, U. (2000). Scratch-resistant conductive coatings. *Google Patents.*

121 Jonas, F., Krafft, W., Muys, B. (1995). *In poly (3, 4-ethylenedioxythiophene): conductive coatings, technical applications and properties. Macromol. Symp.* Wiley Online Library; pp 169–173.

122 Lefebvre, M., Qi, Z., Rana, D., and Pickup, P.G. (1999). Chemical synthesis, characterization, and electrochemical studies of poly (3, 4-ethylenedioxythiophene)/poly (styrene-4-sulfonate) composites. *Chem. Mater. 11*: 262–268.

123 Zhang, F., Johansson, M., Andersson, M.R. et al. (2002). Polymer photovoltaic cells with conducting polymer anodes. *Adv. Mater. 14*: 662–665.

124 Ouyang, J., Chu, C.W., Chen, F.C. et al. (2004). Polymer optoelectronic devices with high-conductivity poly(3,4-ethylenedioxythiophene) anodes. *J. Macromol. Sci., Part A 41*: 1497–1511.

125 Ouyang, J., Chu, C.W., Chen, F.C. et al. (2005). High-conductivity poly (3, 4-ethylenedioxythiophene): poly (styrene sulfonate) film and its application in polymer optoelectronic devices. *Adv. Funct. Mater. 15*: 203–208.

126 Admassie, S., Zhang, F., Manoj, A.G. et al. (2006). A polymer photodiode using vapour-phase polymerized PEDOT as an anode. *Sol. Energy Mater. Sol. Cells 90*: 133–141.

127 Dong, Q., Zhou, Y., Pei, J. et al. (2010). All-spin-coating vacuum-free processed semi-transparent inverted polymer solar cells with PEDOT:PSS anode and PAH-D interfacial layer. *Org. Electron. 11*: 1327–1331.

128 Zhou, Y., Cheun, H., Choi, S. et al. (2010). Indium tin oxide-free and metal-free semitransparent organic solar cells. *Appl. Phys. Lett. 97*: 153304.

129 Zhou, Y., Cheun, H., Choi, S. et al. (2011). Optimization of a polymer top electrode for inverted semitransparent organic solar cells. *Org. Electron. 12*: 827–831.

130 Song, W., Fan, X., Xu, B. et al. (2018). All-solution-processed metal-oxide-free flexible organic solar cells with over 10% efficiency. *Adv. Mater. 30*: e1800075.

131 Gupta, D., Wienk, M.M., and Janssen, R.A.J. (2013). Efficient polymer solar cells on opaque substrates with a laminated PEDOT:PSS top electrode. *Adv. Energy Mater. 3*: 782–787.

132 Zhou, Y., Khan, T.M., Shim, J.W. et al. (2014). All-plastic solar cells with a high photovoltaic dynamic range. *J. Mater. Chem. A 2*: 3492.

133 Dou, L., You, J., Yang, J. et al. (2012). Tandem polymer solar cells featuring a spectrally matched low-bandgap polymer. *Nat. Photonics 6*: 180–185.

134 Zhou, Y., Fuentes-Hernandez, C., Shim, J.W. et al. (2012). High performance polymeric charge recombination layer for organic tandem solar cells. *Energy Environ. Sci. 5*: 9827–9832.

135 Tong, J., Xiong, S., Zhou, Y. et al. (2016). Flexible all-solution-processed all-plastic multijunction solar cells for powering electronic devices. *Mater. Horiz. 3*: 452–459.

136 Roman, L.S., Berggren, M., and Inganäs, O. (1999). Polymer diodes with high rectification. *Appl. Phys. Lett. 75*: 3557–3559.

137 Zhang, F.L., Gadisa, A., Inganäs, O. et al. (2004). Influence of buffer layers on the performance of polymer solar cells. *Appl. Phys. Lett. 84*: 3906–3908.

138 Baierl, D., Fabel, B., Gabos, P. et al. (2010). Solution-processable inverted organic photodetectors using oxygen plasma treatment. *Org. Electron. 11*: 1199–1206.

139 Savva, A., Neophytou, M., Koutsides, C. et al. (2013). Synergistic effects of buffer layer processing additives for enhanced hole carrier selectivity in inverted organic photovoltaics. *Org. Electron. 14*: 3123–3130.

140 Jiang, F.-X., Xu, J.-K., Lu, B.-Y. et al. (2008). Condensed matter: electronic structure, electrical, magnetic, and optical properties: Thermoelectric performance of poly (3, 4-ethylenedioxythiophene): poly (styrenesulfonate). *Chin. Phys. Lett. 25*: 2202–2205.

141 Li, Y., Du, Y., Dou, Y. et al. (2017). PEDOT-based thermoelectric nanocomposites – a mini-review. *Synth. Met. 226*: 119–128.

142 Li, Z., Sun, H., Hsiao, C.-L. et al. (2018). A free-standing high-output power density thermoelectric device based on structure-ordered PEDOT:PSS. *Adv. Electron. Mater. 4*: 1700496.

143 Yue, R. and Xu, J. (2012) Poly(3,4-ethylenedioxythiophene) as promising organic thermoelectric materials: a mini-review. *Synth. Met. 162*: 912–917.

144 Feng-Xing, J., Jing-Kun, X., Bao-Yang, L. et al. (2008). Thermoelectric performance of poly(3,4-ethylenedioxythiophene): poly(styrenesulfonate). *Chin. Phys. Lett. 25*: 2202–2205.

145 Fan, Z., Li, P., Du, D., and Ouyang, J. (2017). Significantly enhanced thermoelectric properties of PEDOT:PSS films through sequential post-treatments with common acids and bases. *Adv. Energy Mater. 7*: 1602116.

146 Dong, X., Xiong, S., Luo, B. et al. (2018). Flexible and transparent organic-inorganic hybrid thermoelectric modules. *ACS Appl. Mater. Interfaces 10*: 26687–26693.

147 Bae, E.J., Kang, Y.H., Jang, K.S., and Cho, S.Y. (2016). Enhancement of thermoelectric properties of PEDOT:PSS and tellurium-PEDOT:PSS hybrid composites by simple chemical treatment. *Sci. Rep. 6*: 18805.

148 Bubnova, O., Khan, Z.U. Malti, A. et al. (2011). Optimization of the thermoelectric figure of merit in the conducting polymer poly(3,4-ethylenedioxythiophene). *Nat. Mater. 10*: 429–433.

149 Kim, G.H., Shao, L., Zhang, K., and Pipe, K.P. (2013). Engineered doping of organic semiconductors for enhanced thermoelectric efficiency. *Nat. Mater. 12*: 719–723.

150 Bahk, J.-H., Fang, H., Yazawa, K., and Shakouri, A. (2015). Flexible thermoelectric materials and device optimization for wearable energy harvesting. *J. Mater. Chem. C 3*: 10362–10374.

151 Søndergaard, R., Hösel, M., Angmo, D. et al. (2012). Roll-to-roll fabrication of polymer solar cells. *Mater. Today 15*: 36–49.

152 Xi, X., Matijasevic, G.S., Brandt, L., and Ha, L. (2000). Process for producing high performance thermoelectric modules. *Google Patents*.

153 Søndergaard, R.R., Hösel, M., Espinosa, N. et al. (2013). Practical evaluation of organic polymer thermoelectrics by large-area R2R processing on flexible substrates. *Energy Environ. Sci. 1*: 81–88.

3

Flexible Thermoelectrics Based on Poly(3,4-Ethylenedioxythiophene)

Ming Hui Chua, Ady Suwardi, and Jianwei Xu

*Institute of Materials Research and Engineering (IMRE), Agency for Science, Research and Technology (A*STAR), Singapore, Singapore*

3.1 Introduction

The emergence of flexible electronics technologies (also known as soft electronics) in the recent decade has ushered in many exciting opportunities and possibilities for electronic materials applications. Notably, flexible electronics technologies have been rigorously employed in the large-scale fabrication of light-harvesting photovoltaic sheets, organic light-emitting diode, electrochromic textiles and coatings, artificial skin, and other medical wearable devices. With this, traditionally hard, heavy, bulky and expensive devices, and generators are steadily making way for new generations of flexible, lightweight, and potentially low-cost alternatives, which bring about great convenience to users. In general, advances in flexible electronics are made possible by two technical aspects: (i) development of flexible lightweight functional materials for key components of devices; and (ii) methods, techniques, and platforms for device fabrication that allow for the functional materials to be appropriately applied and integrated. For the former, materials such as organic polymers and their composites are being developed with the aim to match the performance of their traditional heavier and more rigid counterparts. For the latter, innovative strategies are conceived to engineer flexible devices, which include (i) vacuum-based microfabrication such as lithographic patterning and undercut etching; (ii) solution processing of materials (such as ink-jet printing) onto soft and

Flexible Thermoelectric Polymers and Systems, First Edition. Edited by Jianyong Ouyang.

flexible substrates; (iii) aerogel and hydrogel systems; (iv) use of microfluidics, and (v) fibre-based textile electronics [1–6]. These intensive research efforts invested have led to great advances in the field of flexible electronics with countless scientific reports, numerous device prototypes, as well as several commercialized technologies [4, 7].

Amongst flexible electronics, in addition to attractive features such as light weightiness, ease and comfort of use and wearing, wearable devices have received immense attention because of the ease to collect a large number of users' real-time health data to better protect their health. They may be simply a bracelet or a wrist-band, a skin-patch, an implant, or even a tattoo printed directly on skin [8]. For practicality, wearable devices are lightweight, highly stretchable, and can conform to shape of the body parts. Coupled with the emergence of Internet of Things (IoT), flexible wearable devices promise convenience by facilitating medical diagnosis and allowing personalised from short- to long-term health and fitness monitoring and analysis, which traditionally requires users to manually perform biological laboratory tests and utilize immobile medical equipment and machines housed only in medical institutions. In addition, analysis from these wearable devices is noninvasive and is able to be engineered to measure a range of health and fitness indicators and physiological parameters. These include body temperature, blood pressure, heart rate and electrocardiography (ECG), respiration rate, pulse oxygenation, dynamic pressure and strain on joints and body parts, glucose level and biomarkers from sweat and other bodily fluids as well as brainwave activities [7, 9–12].

One key aspect of wearable device engineering is the power source. Ideally, devices should run on very little energy, which could be provided from conventional replaceable batteries, or simple charging of built-in capacitors. The key is to allow users to retain their level of comfort and mobility while wearing these devices. However, replaceable batteries or capacitors may still be too large, thick, rigid, or noncompliant to be incorporated into smaller, thinner, and more flexible wearable devices. Alternative power sources have been explored and incorporated into the fabrication of thin flexible wearable devices. From the external environment, energy can be harvested from light and radio-frequency electromagnetic waves via photovoltaics and antennas, respectively. From the user's body, energy can also be harvested from body motion via the use of piezoelectrics [13], as well as from body heat via thermoelectrics. In particular, incorporation of thermoelectric generator (TEG) as the power source into a flexible wearable device has been well explored. This is because the operability of the device requires neither an external power source nor dependence of external environment or user's physical activities, and hence it is considered "self-powered" because temperature difference between the body (hot surface) and the external ambient environment (cold surface) is sufficiently high to produce energy to power the entire device.

To realise the widespread use of TEG to power flexible wearables and electronics, intensive research efforts have been invested into flexible TE materials development and engineering of flexible TEGs. Traditional TE materials such as Bi_2Te_3 and Sb_2Te_3 are hard, rigid, and brittle inorganic alloy-based, and possess outstanding TE performance. On the other hand, organic conjugated polymers, such as polyaniline (PANI) and poly(3,4-ethylenedioxythiophene) (PEDOT) are lightweight and flexible, which serve as promising materials compatible for flexible TE electronics and wearables, though they possess poorer TE performance than their inorganic counterparts. Therefore, research in TE materials development includes the design and synthesis of novel conjugated polymers, as well as organic–inorganic hybrid composites materials with the aim of developing TE materials with comparable performance to traditional inorganic TE materials [14–17]. Similarly, the development of flexible TEG is considered a milestone compared to traditional ones [18]. Flexible TEGs can be thin film- or fibre-based, both of which are lightweight, bendable, deformable, and possess a large surface area, and potentially low-cost of production [10, 18–21]. Amongst all reported and commercially available conjugated polymers, PEDOT appears to be the most promising [22–24]. PEDOT often exists as a polymer mixture with polystyrene sulfonate (PSS) in aqueous solution. PEDOT:PSS often serve as the benchmark for organic TE materials with a ZT value of ~0.42, providing an excellent candidate for flexible TE modules. This chapter will summarize the latest development of PEDOT:PSS on the flexible thermoelectrics.

3.2 TE Materials and Devices

3.2.1 Fundamental Principles and Theory of Thermoelectrics

In the last few decades, although technological advancements come in leaps and bounds, there is no significant change in the way that we utilize energy. For instance, in a typical combustion, only about 34% energy is efficiently utilized, while the rest is wasted as heat. This can be partially associated with the fundamental law of physics (i.e. Carnot efficiency) which sets the upper limit efficiency of any energy conversion [25, 26]. On the other hand, more concerted efforts can be dedicated to harvesting the wasted heat and converting it back into useful energy. The ability to convert heat flow into electricity makes TE materials a perfect candidate for this purpose. Thermoelectrics was first discovered in 1821 by German scientist Thomas Johann Seebeck, who discovered the Seebeck effect (potential difference across two junctions with different temperatures). Subsequently, in 1834, French physicist Jean Charles Athanase Peltier discovered the Peltier effect (temperature difference as a result of a current flow), which is the converse effect of Seebeck [27, 28]. Overall, TE materials can be used to serve dual purposes: as a power generator via the Seebeck effect as well as a cooling device via the Peltier effect. The performance metric of a TE material can be characterized by:

$$zT = \frac{S^2\sigma}{\kappa}T \tag{3.1}$$

where S is the Seebeck coefficient, σ is the electrical conductivity, and κ is the thermal conductivity. Seebeck coefficient represents voltage per unit temperature difference. In principle, to obtain a high ZT value, high Seebeck coefficient, high electrical conductivity, and low thermal conductivity are usually essential. Nevertheless, it is challenging to obtain the above combinations due to the interdependencies between these parameters via carrier concentration, which is expressed as:

$$S = \frac{8\pi^2 k_B^2 \mathrm{T}}{3eh^2} m_S^* \left(\frac{\pi}{3n}\right)^{2/3} \tag{3.2}$$

$$\sigma = ne\mu \tag{3.3}$$

Therefore, the carrier concentration (n) always needs to be optimized to get optimum S and σ. For the most inorganic thermoelectrics, n typically ranges from 10^{19} to 10^{21} cm^{-3} whereas for organics materials, n is relatively hard to be quantified and remains a state-of-art challenge in the community (measuring n is especially challenging for materials with a low carrier mobility) [29–33]. Besides n, other strategies (beyond the scope of this report) to synergistically improve S, σ, and κ have been the focus of ongoing researches for many years [34–41]. Machine learning and data driven approach have also been used for exploring high performance thermoelectrics [42–44]. Furthermore, thermal conductivity is also partly dependent on electrical conductivity due to electronic contribution to the total thermal conductivity. To the same end, the relatively low thermal conductivity (partly due to low electrical conductivity) in organic thermoelectrics has been a major advantage as compared to its inorganic ones. Traditionally, the popular way to fabricate inorganic bulk thermoelectrics is by spark plasma sintering (SPS) [45–50]. On a device level, the efficiency not only depends on ZT, but is also limited by Carnot efficiency. Therefore, the efficiency can be expressed as:

$$\eta = \frac{\Delta T}{T_{\text{hot}}} \frac{\sqrt{1 + zT} - 1}{\sqrt{1 + zT} + \frac{T_{\text{cold}}}{T_{\text{hot}}}} \tag{3.4}$$

In equation (3.4), $\Delta T/T_{\text{hot}}$ represents Carnot efficiency, which sets the upper theoretical limit of power generation efficiency. Similar to other heat engines, it is beneficial for TE devices to operate at a high temperature gradient (ΔT) to leverage on the benefit of having high Carnot efficiency. In addition, a high ZT is also desirable to achieve a high device efficiency.

Thermoelectrics found its first application as the radioisotope TEG in space used by the National Aeronautics and Space Administration in 1976 [25, 51]. Other uses including solid-state cooling and temperature sensors have also been commercialized. In recent years, TEGs have been touted for various applications, especially in the flexible and wearable electronics areas such as wrist watches [52–64]. Most of the flexible TEGs are made of PEDOT-based [55, 65–76]. In comparison to other power generation means, thermoelectrics is attractive for wearables due to the constant heat that human body is producing. For instance, the average power dissipated by a resting human body is 116 W. By assuming human body temperature of 37 °C and ambient temperature of −10 °C, the maximum waste power that can be recovered as high as 17.6 W [77]. This number drops to 0.75 W when the ambient temperature increases to 35 °C. Thus, depending on temperature difference, the power that can be harvested from human body ranges from below 1 W to about 10 W. By using simple assumption that the wearable TE device covers only 1% of heat produced by human body, tens of mW can still be harvested, which is sufficient to power small, low power personal electronics such as smart watch, or even small-scale health monitoring electronics.

3.2.2 PEDOT and Its Composites as TE Materials

Conjugated polymers are organic macromolecules containing repeating aromatic or vinylic monomeric moieties with overlapping p orbitals, which create a system of delocalised π electrons along the polymeric backbone. Conjugated polymers exhibit interesting optical and electronic properties. Most conjugated polymers are semi-conductors, and upon doping, they can achieve charge and electrical conductivities, in some cases, comparable to metals. There are also some polymers that are inherently conducting. This makes conjugated polymers potentially good substitutes to metallic substances, and thus are widely synthesized and investigated for several important applications, such as organic photovoltaics (OPV), organic field effect transistors (OFET), organic light-emitting diode (OLED), electrochromic (EC) devices, and more recently, thermoselectrics. In contrast with traditional inorganic-based TE materials, which are hard, rigid, expensive, and possibly toxic, conjugated polymers are lightweight, flexible, relative low-cost to synthesize, low toxicity, and good processability. With appropriate modifications to the chemical structure of the polymeric backbone, big improvements can be made to the solubility of conjugated polymers, thus making them solution processable, and hence extremely useful for fabrication of thin-film or flexible devices.

Many conjugated polymers and their composites including PEDOT, PANI, poly(3-hexylthiophene) (P3HT), and poly(pyrrole) (PPy) [16, 78–84] have been studied for TE applications. Amongst these, PEDOT serves as the benchmark for organic TE materials due to its high electrical conductivity, low thermal conductivity, light-weightiness, and flexibility. In addition, upon appropriate doping or posttreatment, PEDOT appears optically transparent and can achieve enhanced electrical conductivity and remarkably large Seebeck coefficients [22–24, 85]. Conducting PEDOT is commercially available in an aqueous-dispersed form with hydrophilic PSS. The chemical structure of PEDOT:PSS is shown in Scheme 3.1. Ionic interactions exist between negatively charged sulfonate groups of PSS and oxidised radical cationic PEDOT, thus rendering it water dispersible.

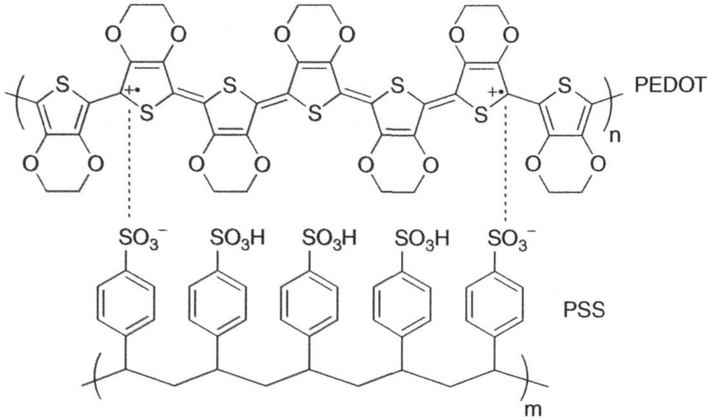

Scheme 3.1 Chemical structure of PEDOT:PSS.

Although PEDOT:PSS is highly processable, the presence of PSS affects chain alignment of PEDOT, and subsequently electrical conductivity. Pristine PEDOT:PSS has an electrical conductivity of less than 1 $S cm^{-1}$. In addition, the Seebeck coefficient of pristine PEDOT:PSS is only $12-16$ $\mu V K^{-1}$ [24, 86]. Intensive research efforts have been invested to improve the TE properties of PEDOT:PSS by preparing composites with different additives or applying different posttreatment. Additives may be mixed or homogenously dispersed into aqueous suspension of PEDOT:PSS, followed by film-forming process. Casted PEDOT:PSS films might have undergone different kinds of posttreatments, such as solvent/solution treatment (dipping or washing films with chemical solutions), and vapor treatment (exposing films to chemical fumes). The rationale of these treatment methods is to increase charge carrier transport by reorientating PEDOT polymer chains and modulating morphology or removing excess insulating PSS that is detrimental to the electrical conductivity of PEDOT.

About 33% PEDOT is already doped in pristine PEDOT:PSS, corresponding to a carrier concentration of $10^{21}-10^{22}$ cm^{-3} [22]. This doping level can be controlled via secondary doping by mixing PEDOT:PSS with a secondary dopant in aqueous suspension, or post-treatment of casted PEDOT:PSS films. Commonly used secondary dopants include polar organic solvents, such as dimethylsulfoxide (DMSO), ethylene glycol (EG), methanol (MeOH), and *N,N*-dimethylformamide (DMF); [67, 87–89] inorganic salts, such as $CuCl_2$ and $ZnCl_2$ [61, 90], as well as various organic and inorganic acids, such as trifluoroacetic acid (TFA), formic acid (FA), and sulfuric acid (H_2SO_4) [87, 91]. These secondary dopants may be solely added, in binary or combination [92–95]. Through many studies, it was understood that the addition of or treatment with polar organic solvents can weaken Coulombic interactions between PEDOT and PSS via screening effect, which leads to phase segregation between the two polymers, thereby causing morphological change of the doped films [96–98]. Similarly, the introduction of organic salts together with organic solvents was also found to create a synergetic effect in screening Coulombic interactions between PEDOT and PSS, facilitating phase segregation of PSS from PEDOT, and inducing conformational change of the latter from benzoid to quinoid structure[61]. On the other hand, acid treatment of PEDOT:PSS was found to not only improve film electrical conductivity but also increase Seebeck coefficient, as well. This was reportedly attributed to the removal of PSS from the polymer mixture, as well as morphology change of PEDOT [91, 95].

Chemical de-doping of PEDOT:PSS films is another effective strategy in improving TE properties. In the process of de-doping, the Seebeck coefficient of PEDOT:PSS increases at the expense of electrical conductivity. This is due to the reduction of charge carrier concentration of PEDOT:PSS upon de-doping, which affects the material's density of states and band structure factors that affect the Seebeck coefficient. At an optimal level of doping, the increase in Seebeck coefficient can, more than, offset the decrease in electrical conductivity, thereby resulting in an enhancement in overall material TE performance. Common de-doping methods include treatment of PEDOT:PSS films with reducing agents, such as hydrazine, ammonium formate, sodium borohydride as well as acids and bases [99–102]. To mitigate the reduction in electrical conductivity, PEDOT:PSS may undergo sequential treatments of secondary doping with polar organic solvents, followed by de-doping with reducing agents or acid treatment. The first step increases polymer conductivity by changing PEDOT's conformation and film morphology. In contrast, the second step increases the Seebeck coefficient by reducing the oxidation level of PEDOT [65, 101, 103].

A considerable number of studies on composites preparation from the mixing of PEDOT:PSS with inorganic additives with large Seebeck coefficients at room temperature have been reported. These additives are often traditional TE materials such as Bi_2Te_3, $(Bi,Sb)_2Te_3$, Te, PbTe, SnSe, and black phosphorus [104–110]. Such organic–inorganic hybrid composites can be simply prepared by mixing additives into PEDOT:PSS solution, followed by solution processing to form PEDOT:PSS films. Similarly, PEDOT:PSS-based composites can be prepared with carbon-nanostructure additives such as graphene, reduced graphene oxide (rGO), carbon nanotubes (CNT), and fullerene [111–119]. These carbon-nanostructures contain a network of sp^2-hybridised carbons which can effectively form strong π-π interactions with PEDOT polymer chains, thereby beneficial for enhancement in electrical conductivity of composites. PEDOT:PSS/inorganic composites, as well as doped and posttreated PEDOT:PSS exhibit improved TE properties without compromising on flexibility and lightweight. In majority cases, PEDOT:PSS composites exhibit p-type characteristics, and therefore can be used as a p-type promising material suitable for the fabrication of TE-powered flexible electronics.

3.2.3 General Configuration of TE Devices and Generators

The configuration of TE devices as power generation modules does not differ much from TE-based cooling device. Fundamentally, as the voltage per unit leg is very low even for high Seebeck coefficient material (less than 1 mVK^{-1}), a lot of legs have to be connected together to achieve a useful voltage range to power devices (Figure 3.1).

In general, in order to achieve an optimal TE device, it is important to match p-type and n-type materials' performance. As shown in Figure 3.1,

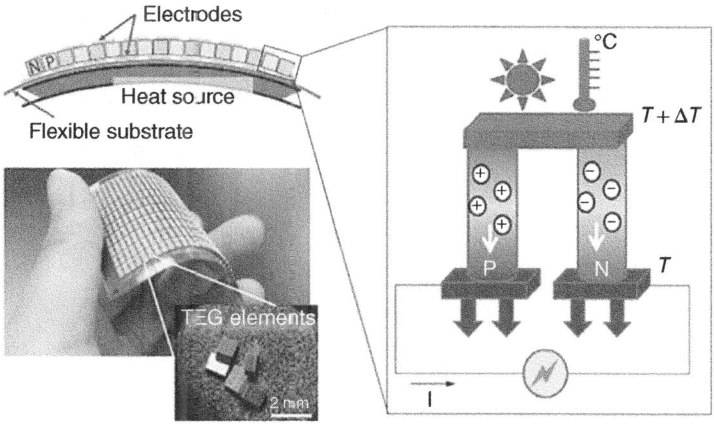

Figure 3.1 Typical flexible TE device configuration showing multiple alternating TE p- and n-type legs connected in series configuration using electrodes and supported by flexible substrates. For optimum performance, the p- and n-type legs must have similar TE performances. *Source:* Figures reproduced with permission from Sugahara et al. [120]. © WILEY-VCH GmbH & Co. KGaA, Weinheim and from Du et al. [77]. © Elsevier.

both the p-leg and n-leg are connected in series via top electrodes (hot side) and bottom electrodes (cold side). Consequently, the temperature gradient ΔT across the hot and cold sides is limited by the leg with a higher thermal conductivity (i.e. lowest ΔT). For instance, if PEDOT:PSS film with thermal conductivity in the range of 0.3–2.2 $\mathrm{W\,mK^{-1}}$ as p-type legs is used to pair with Ag as the n-type legs, the relative ΔT across the device will be low due to heat leakage through the Ag legs with high thermal conductivity of 429 $\mathrm{W\,mK^{-1}}$. The low ΔT ultimately results in a low open circuit voltage and a low power generation efficiency [121]. Therefore, materials with similar thermal conductivity have to be chosen for both p- and n-type legs for an optimum temperature gradient across the hot and cold sides.

In addition to thermal conductivity matching, the geometry of TE legs has to be maximized in terms of the aspect ratio to maximize the temperature gradient as well as external impedance matching. According to Fourier's law, the temperature gradient across hot side and cold side in the presence of heat flow is defined as:

$$Q_h = -\kappa A \frac{\Delta T}{L} \tag{3.5}$$

where Q_h κ, A, L, and ΔT represent heat flow rate, thermal conductivity, cross-sectional area, length, and temperature gradient of the TE legs,

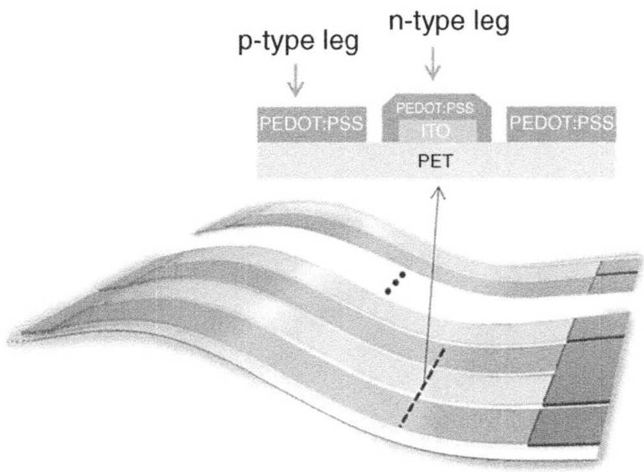

Figure 3.2 Flexible TE thin films with PEDOT:PSS as p-type legs and ITO-doped PEDOT:PSS as n-type legs. The high aspect ratio (length/area) between the cold and hot sides is beneficial for device performance. *Source:* Figure reproduced from Dong et al. [122]. © ACS publications.

respectively. From the above equation, for a given rate of heat flow and thermal conductivity, the ΔT can be maximized by increasing the aspect ratio of L/A (length versus area) of the TE legs. Therefore, to attain a high-performance TE device, it is essential to fabricate legs with a high aspect ratio, as shown in Figure 3.2. Lastly, to maximize the output voltage, it is vital to have the Seebeck coefficients for both the p- and n-type legs because they are connected in series and the resultant thermovoltage adds up between both legs.

3.2.4 Parameters of TE Device and Generator Performances

The performance of a typical TE device can be determined by a set of key parameters namely output voltage, output power density, and device efficiency. In terms of wearable devices, the most important parameter is the output power density, which must be large enough to ensure a reliable and sustainable power for the wearable electronics.

3.2.4.1 Output Voltage

The output voltage of a TE device is defined as:

$$V_{\text{output}} = N\left(S_p - S_n\right)\Delta T \quad (3.6)$$

where N stands for the number of pairs of p-n legs; S_p and S_n represent the Seebeck coefficient of p- and n-type legs (positive for p-type, negative for n-type), respectively; ΔT is the temperature gradient.

On a device level, a few strategies can be employed to maximize the output voltage: increasing the number of legs, choosing TE materials with high Seebeck coefficients and low thermal conductivity to maximize the ΔT. Therefore, depending on the size and geometry of the wearable device as well as the voltage requirements, each of these parameters can be independently adjusted to yield the desired outcomes.

3.2.4.2 Output Power Density

While maximizing the output power is necessary, the ultimate factor that determines the ability of a TE device to power electronic devices is the output power density, which relates to the output voltage via the following equation:

$$P_{\text{density}} = \frac{V_{\text{output}}^{2}}{A\left(R_{TE} + R_L\right)^2} R_L \tag{3.7}$$

where A represents the total cross-sectional area of the device, R_{TE} and R_L represent the total resistance across all TE legs and external load resistance, respectively. To reach the maximum power output, it is of importance to achieve *impedance matching*: a condition when the external load resistance matches the total TE resistance (i.e. $R_L = R_{TE}$). Figure 3.3 illustrates the typical power output curves versus current and voltage in PEDOT:PSS – Te-based

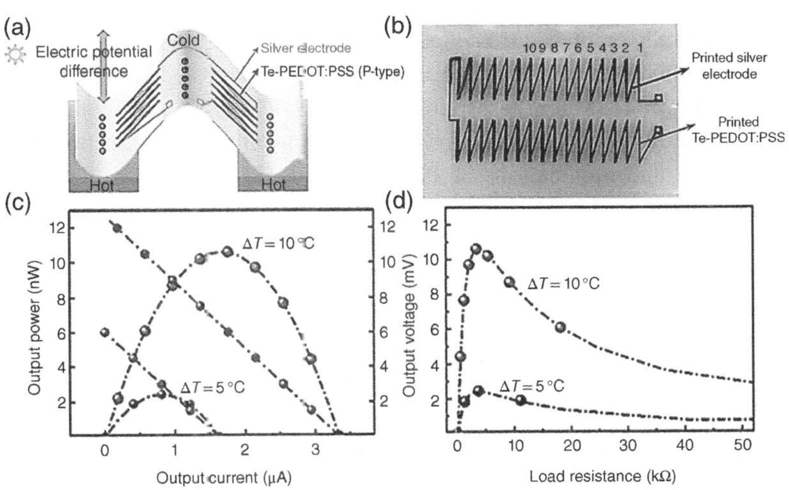

Figure 3.3 (a) Configuration of Te-PEDOT:PSS-based TE devices on flexible substrates. (b) Magnified view showing printed silver electrodes connecting printed Te-PEDOT:PSS TE legs. (c) Typical output characteristic of TE device, showing output power as a function of output current, output voltage, and temperature gradient. (d) The maximum output power happens at the impedance matched condition, when the external load resistance equals to the TE legs resistance. *Source:* Figure reproduced with permission from Bae et al. [123]. © Nature Publishing Group.

flexible TE devices. As shown in Figure 3.3a, PEDOT:PSS and Te are printed onto flexible substrate, with printed silver electrodes as connections. With a temperature gradient across the hot and cold sides, electrical potential differences develop across the device, as illustrated in Figure 3.3b. The maximum output power as a function of temperature gradient is shown in Figure 3.3c, indicating that the higher the ΔT, the higher the output power. Lastly, when the load resistance is equal to the TE resistance (impedance matching), the output power reaches its maximum, as shown in Figure 3.3d.

3.3 PEDOT-Based Flexible TE Materials

Commercially available PEDOT:PSS solutions provide great flexibility for the fabrication process of TE devices, such as spin-coating, drop-casting, ink-jet printing, and roll-to-roll printing. The preparation of PEDOT:PSS composite materials can also be simply performed, in general, by mixing secondary dopants or additives into the solutions, and then a film was casted using a straightforward solution-processing method. Upon drying, the casted film may be peeled off from the surface and can be cut into small pieces for fabrication of TEGs (as the p-type legs). For instance, Caironi et al. reportedly employed inkjet printing to deposit PEDOT:PSS on Kapton substrate, which was then cut into small pieces and annealed at 373 K for an hour to promote water desorption [124]. PEDOT:PSS-based composite materials are highly applicable for flexible electronics. In this section, the preparation and applications of PEDOT:PSS-based flexible TE materials will be summarized.

To enhance the TE properties of PEDOT:PSS, secondary doping and de-doping can be easily carried out by introducing organic polar solvents, small molecules or acids in solution during posttreatment. Zhang et al. reported the preparation PEDOT:PSS-coated paper as an effective and straightforward way toward large-scale production of flexible thermoelectrics [125]. PEDOT:PSS solutions, treated with DMSO and EG, were coated onto paper using Chinese calligraphy (brush-painting). The coated paper exhibited great flexibility and soaking stability in water (Figure 3.4a). More importantly, improved TE property was observed arising from enhancement of the Seebeck coefficient and electrical conductivity, and reduction of thermal conductivity relative to pristine PEDOT:PSS films [125]. Chu et al. prepared PEDOT:PSS flexible bulky papers by vacuum filtration over inorganic filter membrane, which were subsequently treated with secondary dopants EG, PEG, methanol, and FA [94]. Studies show that the film properties of ~100 μm-thick papers (Figure 3.4b) were comparable to that of ~200 nm-thick films, whereas amongst the

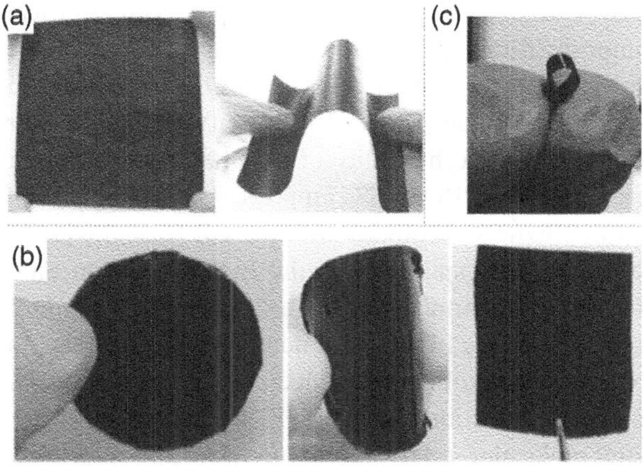

Figure 3.4 (a) Photos of PEDOT:PSS-coated paper prepared by Zhang et al. *Source:* Reproduced with permission from Jiang et al. [125]. © Wiley Periodicals, Inc. (b) Photos of PEDOT:PSS flexible bulky paper prepared by Chu et al. *Source:* Reproduced with permission from Mengistie et al. [94]. © American Chemical Society. (c) Photo of flexible PEDOT:PSS/LiTFSI hybrid films prepared by Liu et al. *Source:* Reproduced with permission Li et al. [126]. © The Royal Society of Chemistry.

secondary dopants, posttreatment with FA yielded the best results, giving rise to an enhanced electrical conductivity of 1900 Scm^{-1}, power factor of 80.6 μWmK^{-2}, compared to 0.3 Scm^{-1} and 0.03 μWmK^{-2}, respectively, for the untreated sample [94]. Ichikawa and Toshima reported preparation of flexible films from PEDOT:PSS/xylitol composites [127]. Interestingly, stretching and posttreatment of the films with DMSO and water resulted in improvement in electrical conductivity, power factor and thus, *ZT* value of the films from 4.03 Scm^{-1}, 0.135 μWmK^{-2}, and $2\ 62\times10^{4}$, to up to 1155 Scm^{-1}, 37.2 μWmK^{-2}, and 5.22×10^{-2}, respectively. This is reportedly due to the improvement in PEDOT chain alignment upon stretching, and the removal of PSS during subsequent posttreatment [127]. More recently, Choi et al. also did a similar study whereby different PEDOT:PSS solutions were mixed with DMSO, EG, PEG, methanol, and H_2SO_4, and then spin-coated onto stretchable PDMS substrate [128]. All the treated samples showed improvement in electrical conductivity and decrease in Seebeck coefficient compared to pristine PEDOT:PSS. Amongst the secondary dopants, PEG was found to increase the stretchability of PEDOT:PSS the most, which contributed to an enhancement of power factor upon stretching, making it a potential candidate for flexible wearable electronics [128].

Introduction of ionic salts can also enhance TE properties of PEDOT:PSS to afford flexible composite materials. Kim et al. reported the preparation of mixed ionic-electronic PEDOT:PSS composite by adding in $CuCl_2$ and then solution-processed to form thin-film devices[90]. Due to Cu^{2+} metal binding onto the PSS, Cl^- channels were formed which enhanced electrical conductivity of the composite film. More interestingly, p-type characteristic of PEDOT:PSS was inverted to n-type in the composite, exhibiting an extremely high negative Seebeck coefficient of -18.2 mV K^{-1} and power factor of 1.7 mW mK^{-2} at a relative humidity of 80% [90]. On the other hand, Liu et al. prepared flexible PEDOT:PSS composites by mixing it with ionic liquid bis(trifluoromethane)sulfonimide lithium salt (LiTFSI) [126]. Flexible films can be obtained by spin-coating the hybrid solution onto Teflon-mold sheet followed by annealing at 130 °C for 20 minutes (Figure 3.4c). The flexible composite film was found to be able to withstand over 10 000 cycles of bending and exhibited enhancement of tensile train by more than 20%. Due to ionic interactions between PEDOT:PSS with LiTFSI, polymer crystallinity and molecular ordering structure were also improved, which contributed to an increase in TE properties as reflected by the enhancement of power factor from 1 to 75 μW mK^{-2} [126].

As previously mentioned, incorporation of hard and rigid traditional inorganic TE materials into PEDOT:PSS can give rise to enhanced TE properties, and meanwhile the softness and flexibility of PEDOT:PSS almost remain intact. Wei et al. prepared flexible PEDOT:PSS composite films with inclusion of plastic reinforcer polyvinyl alcohol (PVA) and $Bi_{0.5}Sb_{1.5}Te_3$. The former was found to improve film-forming ability and fraction strain, whereas the latter contributed to an increase in tensile strength and reduction in fracture strain [129].

As shown in Figure 3.5a, DMSO-treated PEDOT:PSS was first mixed with PVA and $Bi_{0.5}Sb_{1.5}Te_3$, followed by ultrasonic dispersion, the resultant mixture was drop-casted onto glass surface, dried, and finally baked to afford PEDOT:PSS composite film. A record power factor of 47.7 μW mK^{-2} and *ZT* value of 0.05 were achieved from the composite film with a tensile strength of 79.3 MPa and fracture strain of 32.4% [129]. Kar-Narayan et al. prepared PEDOT:PSS-based composites solutions with Bi_2Te_3 and Sb_2Te_3 by in situ mixing two component solutions, and then deposited over flexible polyimide sheet via aerosol jet printing (Figure 3.5b) [130]. As prepared TE modules were found to exhibit great flexibility with repeated flexing to a curvature of 300 m^{-1}. However, electrical resistance of the modules increased slightly upon flexing so that it could not return to original points likely due to the formation of disconnections within the composite materials [130]. More recently, Kim et al. reported the preparation of a series of PEDOT:PSS/$SnSe_{1-x}Te_x$ nanosheets composites by ultrasonic dispersion of the Li^+-intercalated and exfoliated $SnSe_{1-x}Te_x$ nanosheets into PEDOT:PSS solution, followed by filtration through nylon membrane filter paper (Figure 3.6a and b) [131]. Amongst nanosheets of various Te content added,

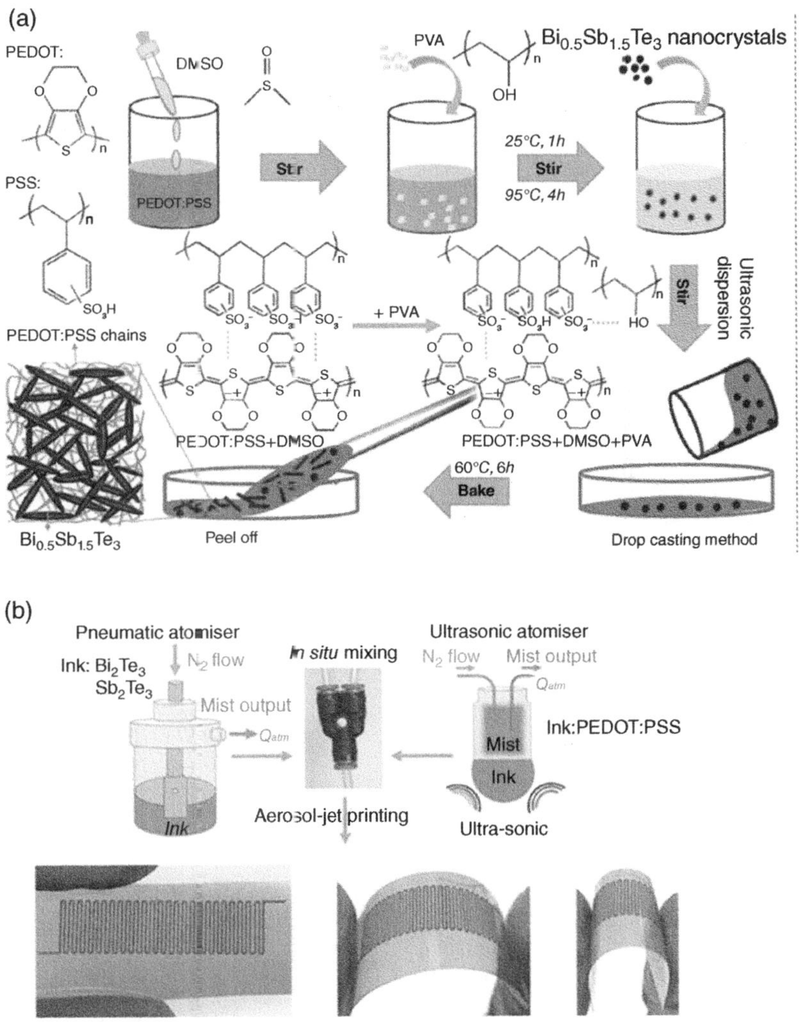

Figure 3.5 (a) Preparation of PEDOT:PSS/PVA/$Bi_{0.5}Sb_{1.5}Te_3$ composite films by Wei et al. *Source:* Reproduced with permission from Zhang et al. [129]. © WILEY-VCH GmbH & Co. KGaA, Weinheim. (b) Aerosol-jet printing of flexible TE films from PEDOT:PSS/Bi_2Te_3 and PEDOT:PSS/Sb_2Te_3 composites prepared by in situ aerosol mixing. *Source:* Reproduced with permission form Ou et al. [130]. © American Chemical Society.

optimal TE property was recorded for PEDOT:PSS/$SnSe_{0.97}Te_{0.03}$ composites with a power factor of 14.73 $\mu W\,mK^{-2}$. The flexible composite material can also withstand 1000 bending cycles without significant loss in electrical conductivity, Seebeck coefficient, and power factor (Figure 3.6c) [131].

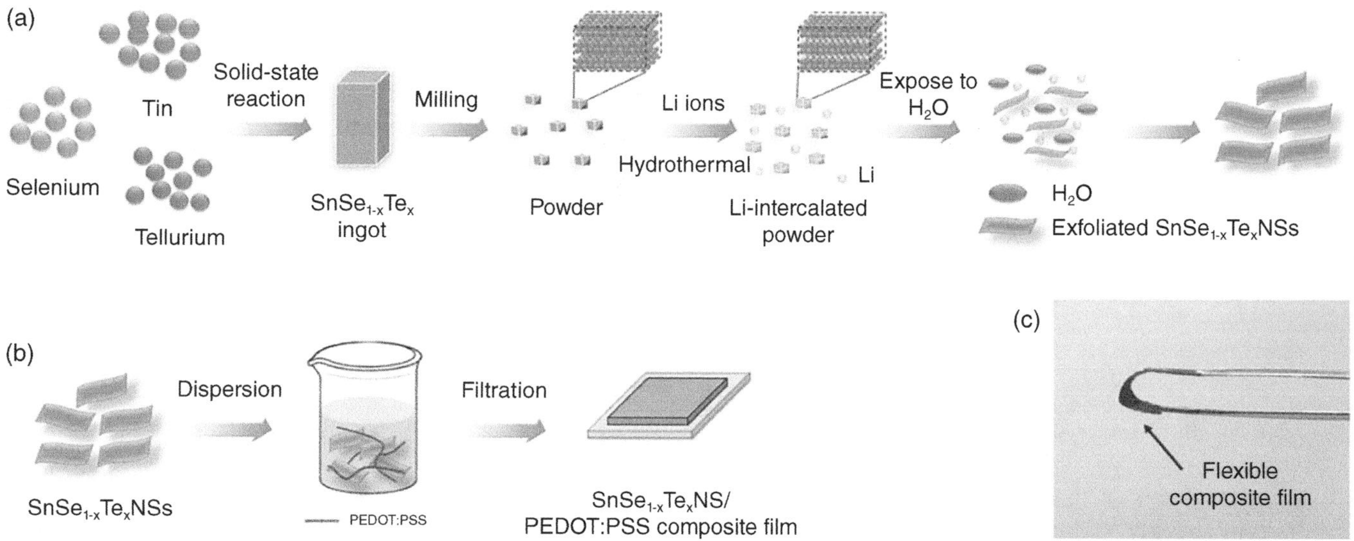

Figure 3.6 (a) Preparation of $SnSe_{1-x}Te_x$ nanosheets. (b) Preparation of PEDOT:PSS/$SnSe_{0.97}Te_{0.03}$ composites. (c) Photo of composite film bending. *Source:* Reproduced with permission from Ju et al. [131]. © Elsevier B.V.

Instead of traditional preparation method of nanocomposite by mixing PEDOT:PSS and inorganic additives to achieve a homogenously dispersed state, Wang et al. reported preparation of flexible PEDOT/Bi_2Te_3 hybrid films by template-assisted nanofabrication [132]. Bi_2Te_3 was deposited over patterned substrates followed by vapor-phase polymerization of PEDOT. At an optimal Bi_2Te_3 composition of 31 vol%, as-prepared flexible hybrid film exhibited an extremely high power factor of ~1,350 $\mu W\,mK^{-2}$ and in-plane *ZT* value of 0.58 at room temperature, and the film could undergo 100 bending cycles with negligible changes to TE properties [132].

PEDOT:PSS composite materials derived from addition of carbon nanostructures such as CNTs are excellent TE materials with lightweight and great flexibility, which was earlier demonstrated by Yu et al. [111]. The high electrical conductivities of these composites were believed to be due to easy transport of electrons across PEDOT–CNT junctions, as well as impedance of CNT's thermal conduction, collectively leading to a power factor as high as ~160 $\mu W\,mK^{-2}$ at room temperature [111]. Subsequently, Xu et al. reported dispersing SWCNTs in PEDOT:PSS and DMSO solution via ultrasonication, followed by vacuum filtration over polyvinylidene fluoride (PVDF) membrane to afford flexible layers of composite material (Figure 3.7a) [114]. At an optimal SWCNT loading of 60 wt%, a maximum *ZT* value of 0.12 was achieved. The composite film can be transferred from PDVF membrane with smooth surface, lightweight, good mechanical flexibility, and foldability, which can withstand rolling-up, bending, twisting, and folding without cracking [114]. Similarly, Yusupov et al. reported the preparation of composite materials by dispersing highly ordered vertically aligned CNT forest (VA-CNTF) into PEDOT:PSS matrix. The use of highly ordered VA-CNTF rather than common disordered CNTs could prevent the formation of agglomerates in composites, contributing to improvement in the material functionalities [133]. EG- and DMSO-posttreated composite films can be easily peeled off from the substrate (Figure 3.7b). The highest power factor obtained from post-treated composites was reported to be 92 $\mu W\,mK^{-2}$ at 415 K [133]. Instead of the conventional solution–dispersion method of mixing additives into PEDOT:PSS, Chen et al. recently reported the preparation of PEDOT/acid-doped SWCNT (a-SWCNT) composites via sonication of PEDOT and crushed a-SWCNT in ethanol solution, whereby PEDOT was synthesized via three-phase interfacial electropolymerization in the presence of PF_6^-, BF_4^-, or ClO_4^- anions, and SWCNT was doped with a mixture of sulfuric acid and nitric acid (Figure 3.8a) [134]. Composite films obtained via vacuum filtration over nylon membrane can be easily bent as shown in Figure 3.8b. More importantly, power factor of the

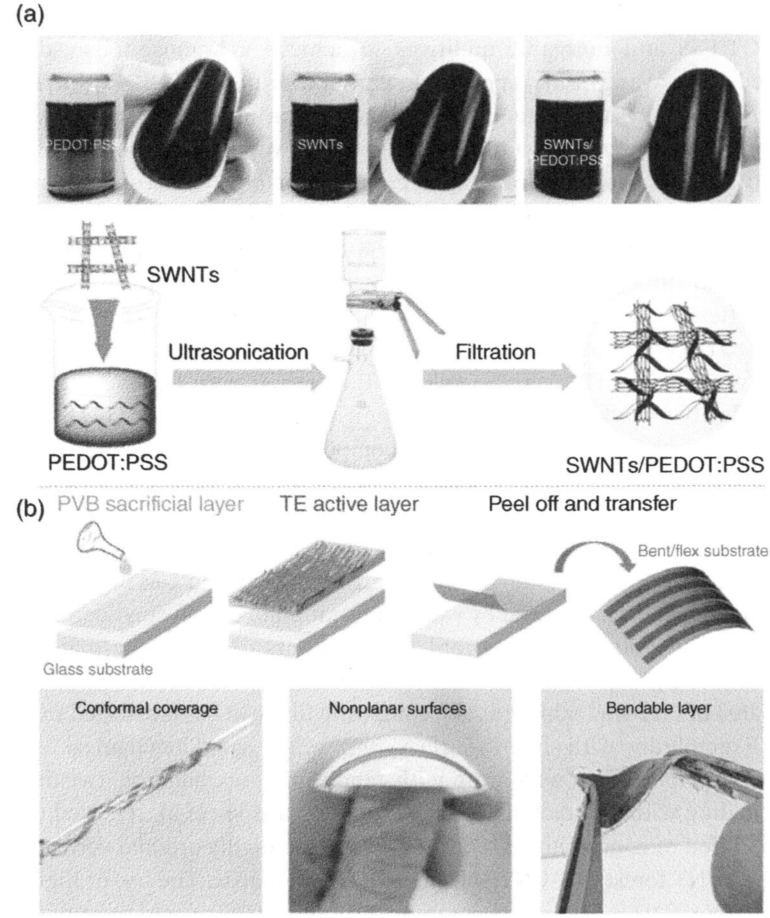

Figure 3.7 (a) Preparation and photos of PEDOT:PSS/SWCNT composite films by Xu et al. *Source:* Reproduced with permission from Jiang et al. [114]. © The Royal Society of Chemistry and the Chinese Chemical Society. (b) Preparation of PEDOT:PSS/VA-CNTF flexible composite films by Yusupov et al. and photos showing the film flexibility and conformal coverage of nonplanar surfaces and irregular shapes. *Source:* Reproduced with permission from Yusupov et al. [133]. © WILEY-VCH GmbH & Co. KGaA, Weinheim.

composites can reach a maximum of 350.0 and 510.2 $\mu W\,m K^{-2}$ at room temperature and 412 K, respectively [134].

As discussed in this section, flexible PEDOT-based TE materials can be easily prepared via straightforward methods such as mixing or posttreatment with a range of dopants and additives from organic polar solvents to acids, CNTs and traditional inorganic TE materials. The preparation and TE

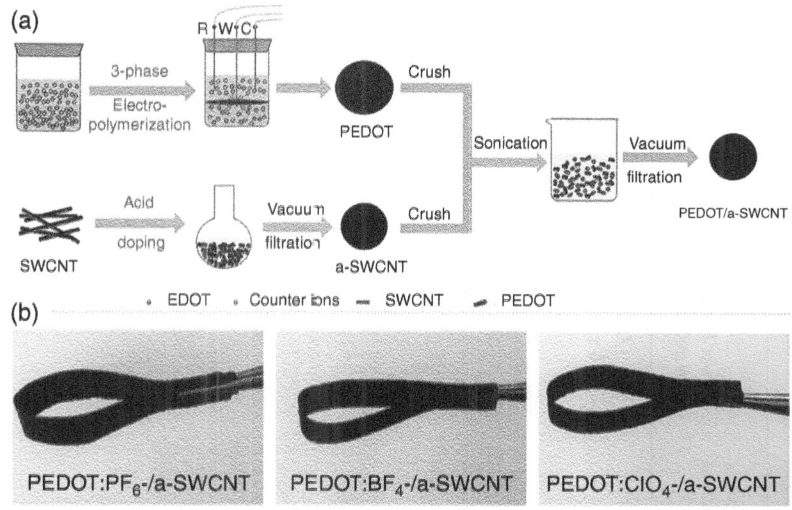

Figure 3.8 (a) Preparation of PEDOT/a-SWCNT composite materials. (b) Photos of PEDOT/a-SWCNT composite films containing different counteranions. *Source:* Reproduced with permission from Fan et al. [134]. © The Royal Society of Chemistry.

properties parameters of the above discussed examples are summarized in Table 3.1. With many outstanding characteristics required for flexible electronics, PEDOT-based composites are potentially used for flexible TEG fabrication and incorporated into body-heat-powered flexible wearable electronics.

3.4 PEDOT:PSS-Based TEGs

In recent years, PEDOT-based TEGs have gained popularity for potential wearable uses due to their ability to be fabricated into any forms or shapes. As a standalone device without the n-type part, the highest reported power density was 56.8 $\mu W cm^{-2}$ with 7.07 mV output voltage across a 100 K temperature gradient [135]. The device consists of five PEDOT:PSS legs on the polyimide connected by silver electrodes, as shown in Figure 3.9. In addition, flexibility test was performed on the entire substrate and device, without notable structural damage. The durability of the device was also tested by exposing to air at 353 K for 4000 hours. On average, before and after the 10 000 bending cycles, the Seebeck coefficient and electrical conductivity were stable at 141 $\mu V K^{-1}$ and 600 $S cm^{-1}$, respectively. Furthermore, both values are nearly constant during the heating cycle, with minor degradation

Table 3.1 Summary of TE properties of flexible PEDOT-based composites.

Primary Material	Additives/ Dopants	Treatment/Device Fabrication	Electrical Conductivity ($S cm^{-1}$)	Seebeck Coefficient ($\mu V K^{-1}$)	Power Factor ($\mu W m K^{-2}$)	*ZT*	*T* (K)	References
PEDOT:PSS	—	Inkjet Printing	11.32 ± 1.57 to 462.86 ± 4.39	~14 to 21	0.23 ± 0.05 to 12.26 ± 0.45	0.0001 to 0.003	298	[124]
PEDOT:PSS	DMSO; EG	Solution coated onto paper	0.20	30.6	3.0	5.5×10^{-3}	300	[125]
PEDOT:PSS	Formic Acid	Preparation of thin film and bulky paper; annealing, then surface treatment	1900	20.6	80.6	0.32	298	[94]
PEDOT:PSS	Xylitol; DMSO or water	Stretching; treated with DMSO or water	1155 ± 115	18.1 ± 1.6	37.2 ± 2.6	0.00522 ± 0.0036	328.2 ± 0.8	[127]
PEDOT:PSS	DMSO; EG; H_2SO_4; MeOH; PEG	Spin-coated onto PDMS then annealed; application of tensile strain	1851 (H_2SO_4); 942 (EG); 891 (DMSO); 1202 (MeOH); 882 (PEG)	18.32 (H_2SO_4); 19.1 (EG); 21.76 (DMSO); 18.7 (MeOH); 19.95 (PEG)	61.18 (H_2SO_4); 13.27 (H_2SO_4 + Strain); 35.1 (PEG); 47.3 (PEG + Strain)	—	298	[128]
PEDOT:PSS	$CuCl_2$	Mixing in solution	—	18200	1700	1.54	298	[90]
PEDOT:PSS	LiTFSI ionic liquid (IL)	IL and PEDOT:PSS stirred and mixed in solution, then dry on Teflon mold	1000	35 (at 200 K)	75	0.05	298	[126]

PEDOT:PSS	$Bi_{0.5}Sb_{1.5}Te_3$; PVA	Solution mixing, ultra-sonication, drop-casting, annealing, then peel off thin film	23	144	47.7	0.05	300	[129]
PEDOT:PSS	Bi_2Te_3; Sb_2Te_3	In situ mixing; aerosol jet printing	~247.3 (Sb_2Te_3)	~33.8 (Sb_2Te_3);	~28.3 (Sb_2Te_3)	—	298	[130]
PEDOT:PSS	$SnSe_{0.97}Te_{0.03}$ nanosheet	Nanosheet powder dispersed in PEDOT:PSS solutions	ca. 18	ca. 90	14.73	—	—	[131]
PEDOT	Bi_2Te_3	Hybrid films prepared via nanosphere lithography and reactive ion etching, over soft substrates	—	Up to ca.185	1350	0.58	298	[132]
PEDOT:PSS	SWCNT	—	1.35×10^3	—	160	—	298	[111]
PEDOT:PSS	SWCNT; DMSO	Mixing by ultra-sonication, then vacuum filtration	200–1400	44.3	105	0.12	—	[114]
PEDOT:PSS	VA-CNTF; DMSO; EG	Dispersion, chemical posttreatment; annealing, then peeled off	Up to 1131	Up to 58.7	Up to 92	—	310; 410	[133]
PEDOT (Prepared by electro-polymerization in presence of ClO_4^-, PF_6^-, or BF_4^- counteranions	a-SWCNT	Mixing and sonication of crushed PEDOT and a-SWCNT in ethanol solution followed by vacuum filtration	4070.8 ± 196.1 (ClO_4^-); 3829.4 ± 87.8 (PF_6^-)	~26 (ClO_4^-); ~30 (PF_6^-)	350.0 ± 47.6 (ClO_4^-); 321.5 ± 35.7 (PF_6^-); 298.9 ± 19.0 (BF_4^-)	—	298	[134]

Figure 3.9 Configuration of five PEDOT:PSS films connected by silver electrodes in series. The device was fabricated in polyimide substrate. *Source:* Figure reproduced with permission from Du et al. [136]. © Royal Society of Chemistry.

in electrical conductivity (17% decrease) after exposing to air for about 3600 hours.

In general, many PEDOT-based TEG devices have been integrated into a wide range of flexible substrates. Although the performance of the device is not as good as the inorganic-based ones, the ability to integrate into flexible substrates has compensated the lack of inherent performance in PEDOT-based TE materials. Table 3.2 summarizes recent PEDOT-based TEG flexible devices and their performances.

Owing to the ability to scale up fabrication process, roll-to-roll printing has been touted as a promising alternative to conventional thin film fabrication method [140]. Zimeng et al. demonstrated a continuous way of fabricating p-type PEDOT:PSS and n-type nitrogen-doped graphene strips using roll-to-roll system. In the process, both PEDOT:PSS and nitrogen-doped graphene were dissolved as inks. The substrate and roller surface of the printing system were treated to improve ink-wettability. By controlling the rotation speed as well as the roll pressure, the inks can be horizontally printed on flexible plastic films. Figure 3.10a shows the schematic of the roll-to-roll printing setup. Continuous roll of plastic substrates was passed through the roll to be first printed with p-type (PEDOT:PSS) strips with a regular spacing. Subsequently, n-type strips were added in between every p-type strip, making alternate variation between p- and n-type strips. Lastly, connector strips that are also made from PEDOT:PSS aqueous solutions were printed using the third roll. As a result, TEGs can be produced in a high-throughput fashion, with alternating p- and n-type junctions that are connected in series, as shown in Figure 3.10b and c. The surface of the substrate as well as roller was not pretreated to improve the ink wettability. As a result, poor wetting can be observed for the final strips, leading to

Table 3.2 PEDOT-based TEG flexible devices and their performances.

Primary Material	Additive/ Dopant	Treatment/Device Fabrication	ΔT (K)	Open Circuit Voltage (mV)	P_{max} ($\mu W cm^{-2}$)	References
PEDOT:PSS	—	Cast solution on Polyimide substrate with Ag electrodes	100	7.07	56.8	[135]
PEDOT:PSS	—	Coated on commercial fabric then connected in series by Ag wires	75.2	4.3	0.026	[137]
PEDOT:PSS	Porous Polyurethane	PEDOT:PSS filled microstructured PU as temperature and pressure sensor	0.1 –100	—	—	[138]
PEDOT:PSS	—	Dispense printing on polyimide substrate with silver paste connection	90	25	100 nW*	[139]
PEDOT:PSS	N doped Graphene	Roll to Roll printing	10	3	0.024	[140]
PEDOT:PSS	Te	Chemical treatment with H_2SO_4 solutions; ECG fabricated by printing process and applied to textile-embedded-type device	10	2	10.59 nW*	[123]
PEDOT:PSS	Ni	Flexible rolled TEG by screen printing on kapton substrate	65	260	46 μW	[141]
PEDOT:PSS	TiS_2	PEDOT:PSS – TiS_2 organic hybrid superlattice films	70	33	250	[142]
PEDOT:PSS	Ag	Inkjet printing on polyimide substrate	5	0.05	0.24 pW*	[121]
PEDOT:PSS	Constantan wires	Cotton fabric coating	74.3	18.7	0.21	[143]
PEDOT:PSS	Te	Fabricated on Polyimide substrates with Ag electrodes	13.4	2.5	—	[144]
PEDOT:PSS	SWNT	Fabricated on flexible plastic substrate	20	—	335 nW*	[145]

(*Continued*)

Table 3.2 (Continued)

Primary Material	Additive/ Dopant	Treatment/Device Fabrication	ΔT (K)	Open Circuit Voltage (mV)	P_{max} ($\mu W\,cm^{-2}$)	References
PEDOT:PSS	AgNPs; Graphene	Printed over knitted textile (polyester/cotton) followed by curing at 120 °C for 30 minutes	100	1.03	—	[146]
PEDOT:PSS	Te, Cu_7Te_4	Eight single legs flexible TE devices	39	32	39.5	[147]
PEDOT:PSS	Te nanorods	PEDOT:PSS-coated Te nanorods prepared by drop-casting followed by H_2SO_4 treatment	40	13.4	57.2	[148]
PEDOT:PSS	SWCNT; Te nanorods	H_2SO_4 treatment	44	5.6	21.4	[149]
PEDOT:PSS	Co/PVDF	Flexible thin films	50	26.4	0.26	[150]
PEDOT:PSS	PEDOT:V_2O_5	Printed on paper	20	14.5	0.266	[151]
PEDOT:PSS	$(Ca_{0.85}Ag_{0.15})_3Co_4O_9$	Flexible thin films	20	0.4	36 nW*	[152]
PEDOT:PSS	ITO	Fabricated on PET substrate	30	5.5	1.75 nW*	[122]
PEDOT	—	PEDOT-coated over textile via vapor-phase polymerization; coated textile alternatively stacked with n-type CNT paper, connected with Cu wires and Ag paste	100	21	0.042	[153]

PEDOT:PSS	AgNPs	AgNPs-coated PEDOT:PSS/PVA fabricated via electrospinning and in situ synthesis (of AgNP)	30.1	3.43	—	[154]
PEDOT:PSS	BiTe-based alloy nanosheet	Composite films adhered on PI substrate and connected in series using Ag paint	47.2	4.8	16.9 nW*	[136]
PEDOT:PSS	—	CH_3NO, H_2SO_4, and $NaBH_4$	12	2.9	1	[155]
PEDOT:PSS	Pristine	Free standing	29	2.0	99	[156]
PEDOT:PSS	Cu_2Se	Nanowire composite thin films	30	15	91	[157]
PEDOT	SWCNT/ Bacterial Cellulose	Aerogels	65.6	6.9	0.014	[158]
PEDOT	Tosylate	Polymer films prepared from EDOT and $Fe(Tos)_3$ via solution casting polymerization	—	—	3.8	[159]
PEDOT	—	Nanowires film	51.6	7.1	157.2 nW*	[160]
PEDOT nanowires	Polydopamine	PEDOT films immersed into PVA-H_2SO_4-polydopamine gel electrolyte; (Supercapacitor)	—	—	1.677	[161]

Note: *Only the total power output was reported in literature and device dimension information is not available.

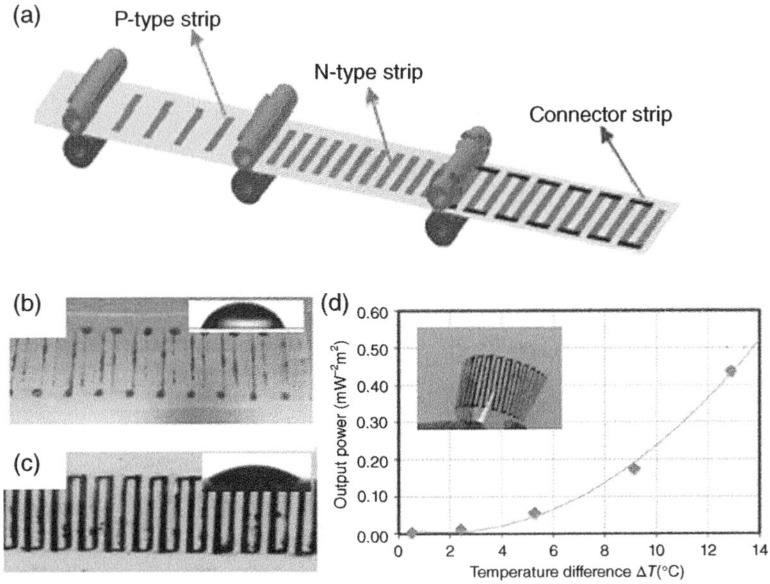

Figure 3.10 (a) Roll-to-roll printing of p- and n-type strips as well as the connector strips, (b) without wettability treatment, (c) with wettability treatment for the substrate and roll, and (d) output power versus temperature difference ΔT of strips. *Source:* Figure reproduced with permission from Zhang et al. [140]. © Elsevier.

discontinuous circuit and hence failed devices. In Figure 3.10c, both substrate and roll have been pretreated to increase the ink wettability, which resulted in consistent deposition of the printed strips. The output power versus temperature difference ΔT is also shown in Figure 3.10d, indicating that the output power of strips is increased with the ΔT.

Besides its compatibility with flexible plastic or polyimide substrate, PEDOT-based TEGs have been demonstrated to be a suitable material for integration into textile fabrics. Minhyun et al. reported their work on using PEDOT:PSS, silver nanoparticles, and graphene inks to fabricate TE devices on polyimide and integrating them into stretchable knitted fabrics [146]. The fabrication process is illustrated in Figure 3.11. In addition, it was demonstrated that the final device could be used as temperature sensors. In the optimized sensor, a TE voltage output of 1.1 mV was generated for a temperature difference of 100 K. A durability of up to 800 cycles and 20% strain was also demonstrated (Figure 3.12). Such structures may be potentially useful for human–machine interface.

As discussed in earlier sections, in order for thermoelectrics to be efficient, minimizing heat flow is essential. Salim et al. developed an intuitive

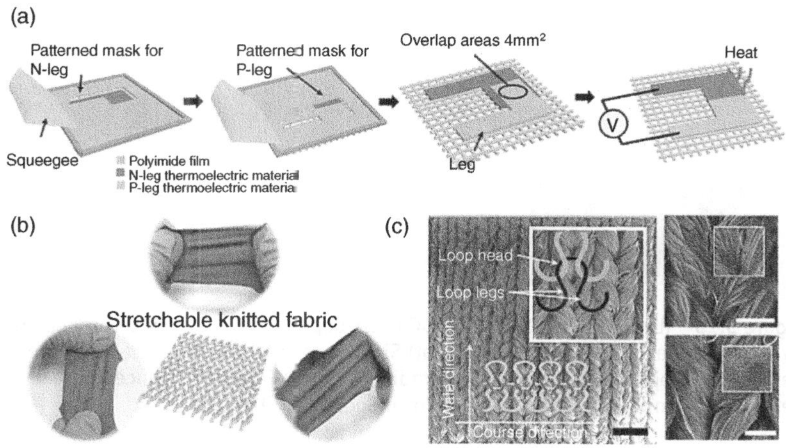

Figure 3.11 Schematic illustration of a stencil-printing method as well as structures of the substrate. (a) Stencil-printing technique employed to fabricate both n-type legs and p-type legs. N-type legs were first printed onto the textile substrate followed by p-type legs with an overlapped area of 4 mm^2. (b) Real image showing stretchable knitted fabrics being stretched in multiple directions. (c) SEM image of the knitted fabric showing loop head and loop legs. Scale bars are 0.2 mm. *Source:* Image reproduced with permission from Jung et al. [146]. © Royal Society of Chemistry.

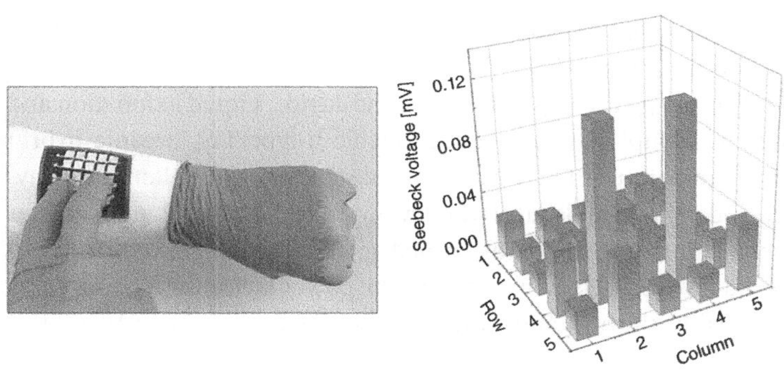

Figure 3.12 Schematic illustration of the final TEG device on knitted fabric for temperature sensor showing corresponding voltage response when two fingers touched the sensors. *Source:* Figure reproduced with permission from Jung et al. [146]. © Royal Society of Chemistry.

technique by fabricating a TEG device onto paper and acting as a substrate. Due to the low thermal conductivity of paper, a high-temperature gradient can be generated across the TEG device with a moderate heat flow [151]. An inkjet printing technology was employed to print hybrid PEDOT:V_2O_5 into a paper substrate. The authors demonstrated that addition of V_2O_5 resulted

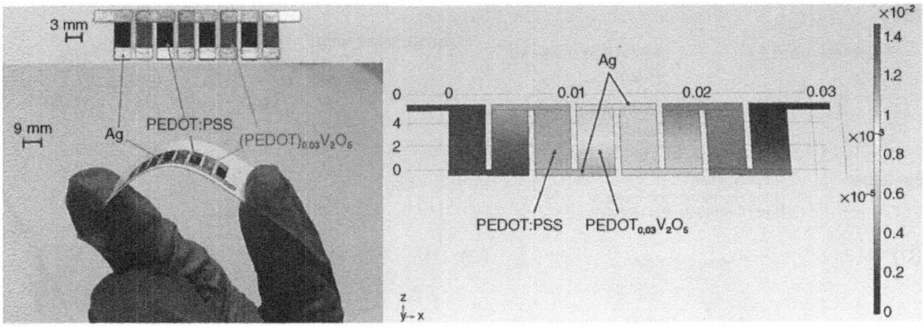

Figure 3.13 (Left) PEDOT:PSS and PEDOT:PSS doped with V_2O_5 device connected by Ag electrodes printed on paper. (Right) Simulated potential gradient when the device was subjected to 20 K thermal gradient. *Source:* Figure reproduced with permission from Ferhat et al. [151]. © Royal Society of Chemistry.

in enhancement of carrier concentration, thus yielding superior TE performance. Furthermore, a numerical model was developed to optimize the geometry of the TEG device. As a result, power density as high as 0.266 μWcm^{-2} was achieved under a temperature gradient of only 20 K. Figure 3.13 shows the real image of the device printed on paper as well as the schematic image with a simulated potential gradient.

In terms of performance, hitherto, the highest power density generated from a PEDOT-based flexible TEG device was achieved by Ruoming et al. [142]. In their work, they utilized TiS_2 as the n-type legs together with PEDOT:PSS as the p-type legs to fabricate the device. Liquid exfoliation and reassembly was employed to produce flexible n-type TiS_2/organic hybrid film for low temperature TEGs. Superlattice structure with alternating layers of TiS_2 and organic molecules was developed to facilitate charge transfer when intercalation complexes was formed, giving rise to high electrical conductivity and low Seebeck. Subsequently, the power factor was enhanced by vacuum annealing the film, and ultimately a power factor of 210 $mWm^{-1} K^{-2}$ was achieved at room temperature. The thermal conductivity of the TiS_2 was reduced by one order of magnitude after intercalation of organic molecules in TiS_2 films. Schematic showing the device, real image of the device as well as the electrical output characteristics is given in Figure 3.14. A comparison of power density, output voltage as well as temperature gradient of PEDOT-based TEG devices is presented in Table 3.1. An interesting trend can be seen, over the years, the main development in this area can be categorized into two directions: demonstrating versatility of substrates used for TEG devices, and improving the output performance of PEDOT-based TEG devices via improving PEDOT:PSS performance or choosing advantageous n-type legs. PEDOT:PSS is on par with TiS_2 in terms of TE characterstic

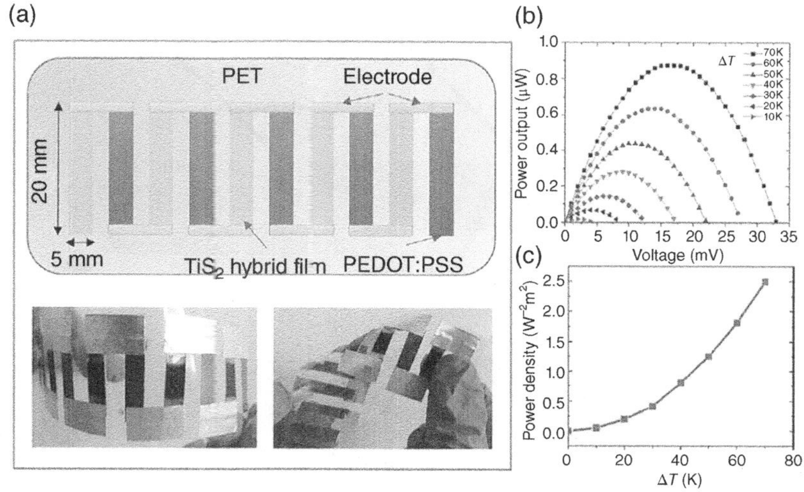

Figure 3.14 (a) Schematic showing p-type legs (PEDOT:PSS) and n-type legs (TiS_2 hybrid film) connected in series by electrodes and deposited on flexible PET substrate (bottom). (b) the power output vs output voltage performance at different temperature gradients. (c) the power output density as a function of temperature gradient. *Source:* Figure reproduced with permission from Tian et al. [142]. © Royal Society of Chemistry.

parameters, exemplifing the importance of not only improving the performance of an individual material (i.e. PEDOT:PSS), but also matching n-type legs with similar TE performance to p-type parts.

Since it is of importance to match the performance of both p- and n-type legs, a facile approach is to use the same TE substance as n- and p-type doping. In a work by Qiufeng et al., PEDOT:PSS-coated Te (PC-Te) nanorod/PEDOT:PSS composite films were prepared by a drop-casting followed by H_2SO_4 treatment to improve the TE properties. As a result of adding PC-Te nanorods, the electrical conductivity and Seebeck coefficients of the composite films were improved dramatically. An optimized power factor of 141.9 $\mu W m K^{-2}$ was obtained for the optimized film at room temperature. A maximum power output density of 57.2 $\mu W cm^{-2}$ was achieved with a temperature difference of 40 K [148].

An alternative approach to avoid the performance discrepancy between p-type legs and n-type legs is by using free-standing, single leg PEDOT:PSS, as demonstrated by Zaifang et al. [156]. In their work, high electrical conductivity of 2500 $S cm^{-1}$ was reported. Therefore, under a small temperature gradient of 29 K, a maximum output power density of 99 ± 18.7 $\mu W cm^{-2}$ was obtained, which to date is the highest ever achieved power density in pristine PEDOT:PSS-based TE devices. In addition, the power output can be

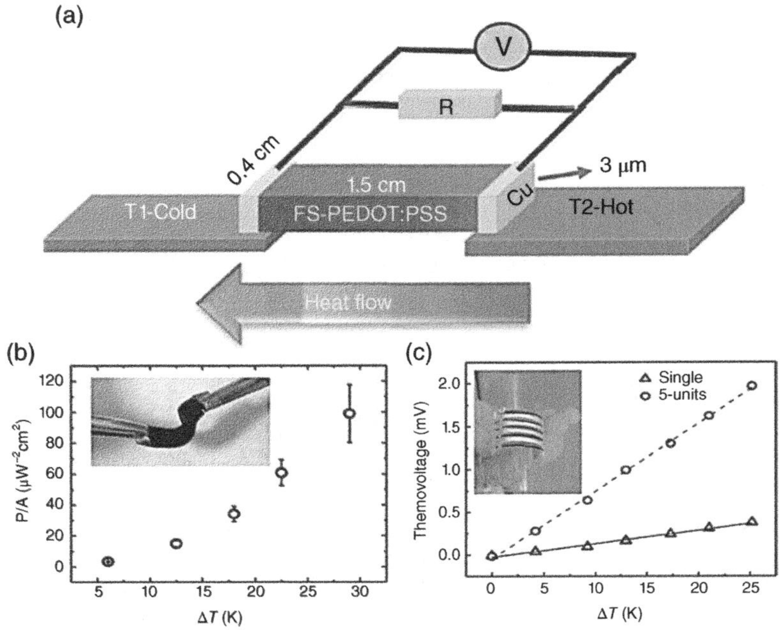

Figure 3.15 (a) Schematic of free standing PEDOT:PSS (b) output power density versus temperature gradient. (inset: free standing PEDOT:PSS connected on both ends with tweezer). (c) Fivefold increase in output voltage by connecting 5 units of TE legs together. *Source:* Figure reproduced with permission from Li et al. [156]. © WILEY-VCH GmbH & Co. KGaA, Weinheim.

controlled by connecting single leg devices in series (i.e. five-fold enhancements in output power was achieved by connecting five devices in series). The simplicity of TEG fabrication using free-standing PEDOT:PSS with good thermal stability of up to 250 °C indicates the tremendous potential to integrate the TEG with textiles for wearable electronics using human body heat. Figure 3.15 illustrates the free-standing single leg PEDOT:PSS with the output characteristics of power density and thermovoltage as a function of temperature gradient for both single leg and five-legs devices.

3.5 Conclusions and Perspectives

Organic-based thermoelectrics has certainly come in leaps and bounds in recent years; however, in terms of development and maturity, it is still lagging behind its inorganic counterparts although there has been tremendous progress in terms of performance (Table 3.1 for PEDOT-based thermoelectrics) as well as devices. In order to enhance *ZT* of PEDOT:PSS, carrier concentration optimization via doping/de-doping, posttreatment, interfacial engineering to achieve energy filtering, as well as improving microstructures and crystallinity have been reported. However, many of these concepts were adopted from inorganic thermoelectrics. Therefore, creative thoughts are required to find out more effective ways to enhance the performance of organic-based thermoelectrics. Moving forward, there are two key areas for the further exploration of this class of material, namely:

1) Up to now, most conducting polymers are p-type materials. Stable and high-performance n-type organic-based TE materials are required to match with its p-type pairs.
2) In general, the geometry of current organic-based devices is entirely based on the way of inorganic TE thermomelectrics. For instance, alternate p- and n-type legs are connected in series. In terms of nature of materials and device structure, such a configuration might not be the most energy-efficient, nor is it a facile way of device fabrication. Therefore, ground-breaking methods are needed for designing new device geometries which particularly cater to organic-based TEG devices.

In terms of devices, thermal resistance between TEGs as well as connecting electrodes, human skin, and air has to be improved to maintain a high temperature gradient to generate sufficient power. More importantly, polymer stability in air and moisture is of paramount importance to ensure the durability of the device. In addition, due to the generally better performance of p-type than n-type materials, the cross-sectional areas of p- and n-type legs should also be optimized. Last but not least, a standard metric for reporting flexible TEG device performance has to be established. To date, the power density per unit area has been the most widely used, albeit some inconstancy (i.e. some papers just reported on the power output, as shown in Table 3.2). In order to reflect the intrinsic performance of the device and also to capture the viability of the TEG to power targeted devices, reporting both critical parameters, the power density and the total power, is highly recommended.

Acknowledgments

The authors acknowledge support from the A*STAR's Science and Engineering Research Council, PHAROS program on Hybrid Thermoelectrics for Ambient Applications: 1527200021 and 1527200019, and Agritech program on Sustainable Hybrid Lighting System for Controlled Environment Agriculture: A19D9a0096.

Conflict of Interests

The authors declare no competing financial interest.

References

1 Bian, J., Zhou, L., Wan, X. et al. (2019). *Adv. Electron. Mater.* 5: 1800900.
2 Yeo, J.C. and Kenry, C.T. (2016). *Lim Lab Chip* 16: 4082.
3 Yuk, H., Lu, B., and Zhao, X. (2019). *Chem. Soc. Rev.* 48: 1642.
4 Weng, W., Chen, P., He, S. et al. (2016). *Angew. Chem. Int. Ed. Eng.* 55: 6140.
5 Kang, J., Tok, J.B.H., and Bao, Z. (2019). *Nat. Electron.* 2: 144.
6 Khan, Z.U., Edberg, J., Hamedi, M.M. et al. (2016). *Adv. Mater.* 28: 4556.
7 Gao, Y., Yu, L., Yeo, J.C., and Lim, C.T. (2019). *Adv. Mater.*: e1902133.
8 Low, Z.W.K., Li, Z., Owh, C. et al. (2019). *Biomater. Sci.*
9 Khan, Y., Ostfeld, A.E., Lochner, C.M. et al. (2016). *Adv. Mater.* 28: 4373.
10 Tao, X. (2019). *Acc. Chem. Res.* 52: 307.
11 Yang, Y. and Gao, W. (2019). *Chem. Soc. Rev.* 48: 1465.
12 Kenry, J.C. and Yeo, C.T. (2016). *Lim Microsyst. Nanoeng.* 2: 16043.
13 Liu, H., Zhong, J., Lee, C. et al. (2018). *Appl. Phys. Rev.* 5: 041306.
14 Jin, H., Li, J., Iocozzia, J. et al. (2019). *Angew. Chem. Int. Ed. Eng.* 58: 15206.
15 Kanahashi, K., Pu, J., and Takenobu, T. (2019). *Adv. Energy Mater.*: 1902842.
16 Russ, B., Glaudell, A., Urban, J.J. et al. (2016). *Nat. Rev. Mater.* 1.
17 Zaia, E.W., Gordon, M.P., Yuan, P., and Urban, J.J. (2019). *Adv. Electron. Mater.* 5: 1800823.
18 Wang, Y., Yang, L., Shi, X.L. et al. (2019). *Adv. Mater.* 31: e1807916.
19 Wang, L. and Zhang, K. (2020). *Energy Environ. Mater.* 3: 67–79. https://doi.org/10.1002/eem2.12045.
20 Zhang, L., Lin, S., Hua, T. et al. (2018). *Adv. Energy Mater.* 8: 1700524.
21 Kishore, R.A., Nozariasbmarz, A., Poudel, B. et al. (2019). *Nat. Commun.* 10: 1765.
22 Fan, Z. and Ouyang, J. (2019). *Adv. Electron. Mater.* 5: 1800769.
23 Song, H., Meng, Q., Lu, Y., and Cai, K. (2019). *Adv. Electron. Mater.* 5: 1800822.
24 Zheng, Y., Zeng, H., Zhu, Q., and Xu, J. (2018). *J. Mater. Chem. C* 6: 8858.
25 Rowe, D.M. (1995). *CRC Handbook of Thermoelectrics*. CRC press.
26 Goldsmid, H.J. (2010). *Introduction to Thermoelectricity*, vol. 121. Springer.
27 Behnia, K. (2015). *Fundamentals of Thermoelectricity*. Oxford: Oxford University Press.
28 Suwardi, A., Prasad, B., Lee, S. et al. (2016). *Nanoscale* 8: 8083.
29 Kang, S.D. and Snyder, G.J. (2017). *Nat. Mater.* 16: 252.

30 Tan, L.P., Sun, T., Fan, S. et al. (2013). *Nano Energy* 2: 4.
31 Suwardi, A., Cao, J., Zhao, Y. et al. (2020). *Mater. Today Phys.* 14: 100239.
32 Suwardi, A., Hu, L., Wang, X. et al. (2020). *ACS Appl. Mater. Interfaces* 12: 9150.
33 Qin, F., Nikolaev, S.A., Suwardi, A. et al. (2020). *Chem. Mater.* 32: 10130.
34 Yan, J., Gorai, P., Ortiz, B. et al. (2015). *Energy Environ. Sci.* 8: 983.
35 Takagiwa, Y., Pei, Y., Pomrehn, G., and Snyder, G.J. (2012). *Appl. Phys. Lett.* 101: 092102.
36 Pei, Y., LaLonde, A.D., Wang, H., and Snyder, G.J. (2012). *Energy Environ. Sci.* 5: 7963.
37 Toberer, E.S., Zevalkink, A., and Snyder, G.J. (2011). *J. Mater. Chem.* 21: 15843.
38 Snyder, G.J. and Toberer, E.S. (2010). Complex thermoelectric materials. In: *Materials for Sustainable Energy: A Collection of Peer-Reviewed Research and Review Articles from Nature Publishing Group* (ed. V. Dusastre), 101–110. World Scientific.
39 Pichanusakorn, P. and Bandaru, P. (2010). *Mater. Sci. Eng. R. Rep.* 67: 19.
40 Kanatzidis, M.G. (2009). *Chem. Mater.* 22: 648.
41 Suwardi, A., Cao, J., Hu, L. et al. (2020). *J. Mater. Chem. A* 8: 18880.
42 Suwardi, A., Bash, D., Ng, H.K. et al. (2019). *J. Mater. Chem. A* 7: 23762.
43 Recatala-Gomez, J., Suwardi, A., Nandhakumar, I. et al. (2020). *ACS Appl. Energy Mater.* 3: 2240.
44 Recatala-Gomez, J., Kumar, P., Suwardi, A. et al. (2020). *Sci. Rep.* 10 (1).
45 Wang, X., Suwardi, A., Zheng, Y. et al. (2020). *ACS Appl. Nano Mater.* 3: 10156.
46 Recatala-Gomez, J., Ng, H.K., Kumar, P. et al. (2020). *ACS Appl. Mater. Interfaces* 12: 33647.
47 Suwardi, A., Lim, S.H., Zheng, Y. et al. (2020). *J. Mater. Chem. C* 8: 16940.
48 Yang, F., Wu, J., Suwardi, A. et al. (2021). *Adv. Mater.* 33: 2004786.
49 Zheng, Y., Xie, H., Zhang, Q. et al. (2020). *ACS Appl. Mater. Interfaces* 12: 36186.
50 Cao, J., Chien, S.W., Tan, X.Y. et al. *ChemNanoMat* n/a.
51 Rowe, D.M. (2018). *CRC Handbook of Thermoelectrics*. CRC Press.
52 Fan, Z. and Ouyang, J. (2019). *Organic Thermoelectric Materials*, vol. 24, 117.
53 Wang, C., Sun, K., Fu, J. et al. (2018). *Adv. Sustain. Syst.* 2: 1800085.
54 Hu, L., Li, M., Yang, K. et al. (2018). *J. Mater. Chem. A* 6: 16583.
55 Yildirim, E., Wu, G., Yong, X. et al. (2018). *J. Mater. Chem. C* 6: 5122.
56 Zhang, S., Fan, Z., Wang, X. et al. (2018). *J. Mater. Chem. A* 6: 7080.
57 Wu, F., Li, P., Sun, K. et al. (2017). *Adv. Electron. Mater.* 3: 1700047.
58 Xia, Y., Fang, J., Li, P. et al. (2017). *ACS Appl. Mater. Interfaces* 9: 19001.
59 Fan, Z., Li, P., Du, D., and Ouyang, J. (2017). *Adv. Energy Mater.* 7: 1602116.

60 Fan, Z., Du, D., Yao, H., and Ouyang, J. (2017). *ACS Appl. Mater. Interfaces* 9: 11732.

61 Fan, Z., Du, D., Yu, Z. et al. (2016). *ACS Appl. Mater. Interfaces* 8: 23204.

62 Guo, C.X., Sun, K., Ouyang, J., and Lu, X. (2015). *Chem. Mater.* 27: 5813.

63 Xia, Y. and Ouyang, J. (2012). *Org. Electron.* 13: 1785.

64 Dinh, K.N., Sun, Y., Pei, Z. et al. (2020). *Small* 16: 1905885.

65 Yemata, T.A., Zheng, Y., Kyaw, A.K.K. et al. (2020). *RSC Adv.* 10: 1786.

66 Yemata, T.A., Kyaw, A.K.K., Zheng, Y. et al. (2019). *Polym. Int.*

67 Yildirim, E., Zhu, Q., Wu, G. et al. (2019). *J. Phys. Chem. C* 123: 9745.

68 Yemata, T.A., Zheng, Y., Kyaw, A.K.K. et al. (2019). *Front. Chem.* 7: 870.

69 Xu, J., Zhu, Q., Yildirim, E. et al. (2019). *Front. Chem.* 7: 783.

70 Kyaw, A.K.K., Yemata, T.A., Wang, X. et al. (2018). *Macromol. Mater. Eng.* 303: 1700429.

71 Wang, X., Kyaw, A.K.K., Yin, C. et al. (2018). *RSC Adv.* 8: 18334.

72 Ling, H., Ding, G., Mandler, D. et al. (2016). *Chem. Commun.* 52: 9379.

73 Ding, G., Cho, C.M., Chen, C. et al. (2013). *Org. Electron.* 14: 2748.

74 Song, J., Jańczewski, D., Ma, Y. et al. (2013). *Eur. Polym. J.* 49: 2477.

75 Ye, T., Wang, X., Li, X. et al. (2017). *J. Mater. Chem. C* 5: 1255.

76 Wang, X., Suwardi, A., Lim, S.L. et al. (2020). *npj Flex. Electron.* 4: 1.

77 Du, Y., Xu, J., Paul, B., and Eklund, P. (2018). *Appl. Mater. Today* 12: 366.

78 Yao, C.J., Zhang, H.L., and Zhang, Q. (2019). *Polymers (Basel)* 11.

79 Zhang, Y., Heo, Y.J., Park, M., and Park, S.J. (2019). *Polymers (Basel)* 11.

80 Yao, H., Fan, Z., Cheng, H. et al. (2018). *Macromol. Rapid Commun.* 39: e1700727.

81 Toshima, N. (2017). *Synth. Met.* 225: 3.

82 Bharti, M., Singh, A., Samanta, S., and Aswal, D.K. (2018). *Prog. Mater. Sci.* 93: 270.

83 Sun, Y., Di, C.-A., Xu, W., and Zhu, D. (2019). *Adv. Electron. Mater.* 5: 1800825.

84 Tang, J., Chen, Y., McCuskey, S.R. et al. (2019). *Adv. Electron. Mater.* 5: 1800943.

85 Petsagkourakis, I., Kim, N., Tybrandt, K. et al. (2019). *Advanced Electronic Materials* 5: 1800918.

86 Xiong, J., Jiang, F., Zhou, W. et al. (2015). *RSC Adv.* 5: 60708.

87 Kyaw, A.K.K., Wong, G.D.H., Yemata, T.A., and Xu, J. (2019). *Org. Electron.* 69: 7.

88 Luo, J., Billep, D., Blaudeck, T. et al. (2014). *J. Appl. Phys.* 115: 054908.

89 Yi, C., Wilhite, A., Zhang, L. et al. (2015). *ACS Appl. Mater. Interfaces* 7: 8984.

90 Kim, B., Hwang, J.U., and Kim, E. (2020). *Energy Environ. Sci.* 13: 859–867. https://doi.org/10.1039/C9EE02399B.

91 Yemata, T.A., Kyaw, A.K.K., Zheng, Y. et al. (2020). *Polym. Int.* 69: 84.

92 Wei, Q., Mukaida, M., Naitoh, Y., and Ishida, T. (2013). *Adv. Mater.* 25: 2831.

93 Wang, X., Meng, F., Tang, H. et al. (2017). *J. Mater. Sci.* 52: 9806.

94 Mengistie, D.A., Chen, C.-H., Boopathi, K.M. et al. (2015). *ACS Appl. Mater. Interfaces* 7: 94.

95 Kumar, S.R.S., Kurra, N., and Alshareef, H.N. (2016). *J. Mater. Chem. C* 4: 215.

96 Ouyang, J., Xu, Q., Chu, C.-W. et al. (2004). *Polymer* 45: 8443.

97 Nardes, A.M., Janssen, R.A.J., and Kemerink, M. (2008). *Adv. Funct. Mater.* 18: 865.

98 Timpanaro, S., Kemerink, M., Touwslager, F.J. et al. (2004). *Chem. Phys. Lett.* 394: 339.

99 Park, H., Lee, S.H., Kim, F.S. et al. (2014). *J. Mater. Chem. A* 2: 6532.

100 Massonnet, N., Carella, A., Jaudouin, O. et al. (2014). *J. Mater. Chem. C* 2: 1278.

101 Lee, S.H., Park, H., Son, W. et al. (2014). *J. Mater. Chem. A* 2: 13380.

102 Yi, C., Zhang, L., Hu, R. et al. (2016). *J. Mater. Chem. A* 4: 12730.

103 Lee, S.H., Park, H., Kim, S. et al. (2014). *J. Mater. Chem. A* 2: 7288.

104 Zhang, B., Sun, J., Katz, H.E. et al. (2010). *ACS Appl. Mater. Interfaces* 2: 3170.

105 Song, H., Liu, C., Zhu, H. et al. (2013). *J. Electron. Mater.* 42: 1268.

106 Du, Y., Cai, K.F., Chen, S. et al. (2014). *ACS Appl. Mater. Interfaces* 6: 5735.

107 Coates, N.E., Yee, S.K., McCulloch, B. et al. (2013). *Adv. Mater.* 25: 1629.

108 Ju, H. and Kim, J. (2016). *ACS Nano* 10: 5730.

109 Lee, J.-Y. and Lin, Y.-J. (2016). *Synth. Met.* 212: 180.

110 Novak, T.G., Shin, H., Kim, J. et al. (2018). *ACS Appl. Mater. Interfaces* 10: 17957.

111 Yu, C., Choi, K., Yin, L., and Grunlan, J.C. (2011). *ACS Nano* 5: 7885.

112 Moriarty, G.P., De, S., King, P.J. et al. (2013). *J. Polym. Sci. B Polym. Phys.* 51: 119.

113 Song, H., Qiu, Y., Wang, Y. et al. (2017). *Compos. Sci. Technol.* 153: 71.

114 Jiang, Q., Lan, X., Liu, C. et al. (2018). *Mater. Chem. Front.* 2: 679.

115 Li, F., Cai, K., Shen, S., and Chen, S. (2014). *Synth. Met.* 197: 58.

116 Yoo, D., Kim, J., and Kim, J.H. (2014). *Nano Res.* 7: 717.

117 Xiong, J., Jiang, F., Shi, H. et al. (2015). *ACS Appl. Mater. Interfaces* 7: 14917.

118 Zhang, K., Zhang, Y., and Wang, S. (2013). *Sci. Rep.* 3: 3448.

119 Li, X., Liang, L., Yang, M. et al. (2016). *Org. Electron.* 38: 200.

120 Sugahara, T., Ekubaru, Y., Nong, N.V. et al. *Adv. Mater. Technol.* 2019 (4): 1800556.

121 Beretta, D., Massetti, M., Lanzani, G., and Caironi, M. (2017). *Rev. Sci. Instrum.* 88: 015103.

122 Dong, X., Xiong, S., Luo, B. et al. (2018). *ACS Appl. Mater. Interfaces* 10: 26687.

123 Bae, E.J., Kang, Y.H., Jang, K.-S., and Cho, S.Y. (2016). *Sci. Rep.* 6: 18805.

124 Beretta, D., Barker, A.J., Maqueira-Albo, I. et al. (2017). *ACS Appl. Mater. Interfaces* 9: 18151.

125 Jiang, Q., Liu, C., Xu, J. et al. (2014). *J. Polym. Sci. B Polym. Phys.* 52: 737.
126 Li, Q., Deng, M., Zhang, S. et al. (2019). *J. Mater. Chem. C* 7: 4374.
127 Ichikawa, S. and Toshima, N. (2015). *Polym. J.* 47: 522.
128 Jeong, M.H., Sanger, A., Kang, S.B. et al. (2018). *J. Mater. Chem. A* 6: 15621.
129 Zhang, T., Li, K., Li, C. et al. (2017). *Adv. Electron. Mater.* 3: 1600554.
130 Ou, C., Sangle, A.L., Datta, A. et al. (2018). *ACS Appl. Mater. Interfaces* 10: 19580.
131 Ju, H., Park, D., Kim, K., and Kim, J. (2019). *J. Alloys Compd.* 792: 638.
132 Wang, L., Zhang, Z., Liu, Y. et al. (2018). *Nat. Commun.* 9: 3817.
133 Yusupov, K., Stumpf, S., You, S. et al. (2018). *Adv. Funct. Mater.* 28: 1801246.
134 Fan, W., Liang, L., Zhang, B. et al. (2019). *J. Mater. Chem. A* 7: 13687.
135 Hokazono, M., Anno, H., and Toshima, N. (2014). *J. Electron. Mater.* 43: 2196.
136 Du, Y., Liu, X., Xu, J., and Shen, S.Z. (2019). *Mater. Chem. Front.* 3: 1328.
137 Du, Y., Cai, K., Chen, S. et al. (2015). *Sci. Rep.* 5: 6411.
138 Zhang, F., Zang, Y., Huang, D. et al. (2015). *Nat. Commun.* 6: 8356.
139 Stepien, L., Roch, A., Schlaier, S. et al. (2016). *Energy Harvesting and Systems*, vol. 3, 101.
140 Zhang, Z., Qiu, J., and Wang, S. (2016). *Manuf. Lett.* 8: 6.
141 Fang, H., Popere, B.C., Thomas, E.M. et al. (2017). *J. Appl. Polym. Sci.* 134.
142 Tian, R., Wan, C., Wang, Y. et al. (2017). *J. Mater. Chem. A* 5: 564.
143 Du, Y., Cai, K., Shen, S. et al. (2017). *RSC Adv.* 7: 43737.
144 Song, H. and Cai, K. (2017). *Energy* 125: 519.
145 Wang, L., Zhang, Z., Geng, L. et al. (2018). *Energy Environ. Sci.* 11: 1307.
146 Jung, M., Jeon, S., and Bae, J. (2018). *RSC Adv.* 8: 39992.
147 Lu, Y., Qiu, Y., Jiang, Q. et al. (2018). *ACS Appl. Mater. Interfaces* 10: 42310.
148 Meng, Q., Jiang, Q., Cai, K., and Chen, L. (2019). *Org. Electron.* 64: 79.
149 Meng, Q., Cai, K., Du, Y., and Chen, L. (2019). *J. Alloys Compd.* 778: 163.
150 Chen, Y., He, M., Tang, J. et al. (2018). *Adv. Electron. Mater.* 4: 1800200.
151 Ferhat, S., Domain, C., Vidal, J. et al. (2018). *Sustain. Energy Fuels* 2: 199.
152 Wang, X., Meng, F., Tang, H. et al. (2018). *Synth. Met.* 235: 42.
153 Jia, Y., Shen, L., Liu, J. et al. (2019). *J. Mater. Chem. C* 7: 3496.
154 Jin, S., Sun, T., Fan, Y. et al. (2019). *Polymer* 167: 102.
155 Xu, S., Hong, M., Shi, X.-L. et al. (2019). *Chem. Mater.* 31: 5238.
156 Li, Z., Sun, H., Hsiao, C.L. et al. (2018). *Adv. Electron. Mater.* 4: 1700496.
157 Lu, Y., Ding, Y., Qiu, Y. et al. (2019). *ACS Appl. Mater. Interfaces* 11: 12819.
158 Jia, F., Wu, R., Liu, C. et al. (2019). *ACS Sustain. Chem. Eng.*
159 Kim, B., Han, M., and Kim, E. (2019). *J. Mater. Chem. A* 7: 2066.
160 Ni, D., Song, H., Chen, Y., and Cai, K. (2019). *Energy* 170: 53.
161 Ni, D., Chen, Y., Song, H. et al. (2019). *J. Mater. Chem. A* 7: 1323.

4

Flexible Thermoelectric Plastic Via Electrochemistry

Fengxing Jiang[1,2], Peipei Liu[2], Baoyang Lu[1], Congcong Liu[1], and Jingkun Xu[1]

[1] *Flexible Electronics Innovation Institute, Jiangxi Science and Technology Normal University, Nanchang, P. R. China*
[2] *Department of Physics, Jiangxi Science and Technology Normal University, Nanchang, P. R. China*

4.1 Introduction

Electrochemical method as a powerful tool can be used for organic synthesis, characterization, and catalysis, sensing, energy storage, etc. through charge transfer, electric energy and chemical energy conversion between organic molecules or catalytic media at the interface of electrode/solution [1]. There are generally two methods for the preparation of conducting polymers (CPs) including chemical and electrochemical methods. Chemical methods can obtain a large number of products by coupling reaction or oxidants in solution at one time, often involving the complex operations and demanding conditions. Also, most of the as-prepared polymers obtained by chemical method are powdery solids, which are insoluble and difficult to process further. This greatly hinders the mechanical and electronic properties of polymer materials and limits their application. In contrast, the electrochemical method has the advantages of generating a free-standing conductive film (doped state) on the electrode in one step without using a catalyst, and the film thickness can be controlled by the amount of electric quantity or the time of the applied voltage. Very thin polymer films can be obtained by electrochemical polymerization, which can be used in many fields. Moreover, the electrochemical method can be used to regulate the doping type and level of polymers by controlling applied voltage.

Flexible Thermoelectric Polymers and Systems, First Edition. Edited by Jianyong Ouyang.

For thermoelectrics, the free-standing films are much desired for the direct application without secondary process such as spin-coating or dip-coating on a substrate. As a promising energy conversion material, organic thermoelectric material has potential application prospects in flexible wearable electronic energy devices due to its light weight, rich resources, simple preparation, safety, reliability and so on. Generally, as a semiconductor material, a thermoelectric material needs to achieve carrier transport through doping. The effective control of carrier concentration and mobility is one of the main ways to optimize the thermoelectric properties of materials. In addition, molecular structure, crystal orientation, and morphology are directly or indirectly related to thermoelectric transport performance. Based on the advantages of electrochemistry, it is one of the effective choices for preparing and regulating the thermoelectric properties of organic thermoelectric materials.

In recent years, electrochemical technology has been used to prepare organic thermoelectric materials and to regulate their regulatory performance. A lot of work has been focused on this regard. This chapter mainly focuses on the electrochemical method preparation, doping principle, performance control, and effect of high-quality conductive polymers.

4.2 Electrochemical Deposition of CPs

Many current works have presented a comprehensive and systematic introduction to electrochemically synthesized CPs from fundamental concepts to practical applications including electrosynthesis with well-designed composition and morphology, and have made outstanding contributions to the development of CPs [2]. Although it turns out that electropolymerization is probably not the technology of choice for mass industrial production, this technology is essential for basic research on monomer polymerization and corresponding polymer properties [2a]. Excluding the general drawbacks of electrochemical preparation produced in rather small amounts, there are several advantages of electrochemical synthesis technique:

1) Low raw material requirements: Electrochemical only requires 10–50 mg of material to perform the corresponding work, which has a positive significance for the full use of drugs and cost reduction. In contrast, chemical polymerization requires a large number of monomers to polymerize, and also requires complex separation and processing before corresponding characterization and application.

2) Rapid analyses: Electrochemical studies can quickly provide information about the properties of electropolymeric materials.
3) Accuracy and precision: Electropolymerization can be performed under the same conditions, allowing comparisons between different systems and facilitating performance screening. In addition, the progress of electrochemical reaction can be effectively controlled to obtain a desired film thickness, doping degree, morphology and so on.

During the electrochemical synthesis of CPs, polythiophene (PTh), polypyrrole (PPy), polyaniline (PANi), and their derivatives have been studied widely. The mechanism of electrochemical polymerization of thiophene, pyrrole, and benzene is still unclear. The radically accepted cationic mechanism is generally accepted. PTh and PPy are taken as an example for the mechanism of electrochemical polymerization, as shown in Scheme 4.1.

Recently, most of the reported research results have excluded CPs formed based on simple chain growth mechanisms. The three main steps involved in electrochemical polymerization are the initial formation of oligomers, nucleation, and growth [2b]. The initial step is the electrochemical oxidation of monomers near the anode to generate free radical cations. Two monomers free radical cations then react with each other to dimerize and remove two protons to form a dimer. The body is reoxidized to form a dimer free radical cation, which is combined with another free radical cation before deprotonation and continued oxidation. This process is repeated to form oligomers. As the oligomer chain length increases, its oxidation potential shifts negatively and the charge delocalization degree in the oligomer is higher than that of the monomer, so the coupling speed between the oligomer radical cation and other radical cations is faster than that of the monomer. The rate of dimerization of bulk free radical cations is high. Owing to the mutual coupling of cationic radicals, the oligomer chain growth takes place first. Once the chain length of the oligomer exceeds the solubility limit in the solvent, deposition and nucleation of the oligomer on the anode will occur. On the electrode, the oligomer chain can continue to grow to form a long-chain polymer, while the polymer conjugated chain formed at the same time is oxidized to form an anion-doped state (conductive state). In the above mechanism, the rate determining step is a dimerization process. As the polymerization deprotonates, the reaction zone near the anode will become acidic.

So far, there is no direct experimental evidence for the other steps except for the oxidation of monomers on the anode to generate free radical cations. In addition, the actual electrochemical polymerization of monomers involves the applied potentials, solvents, electrolytes, temperature,

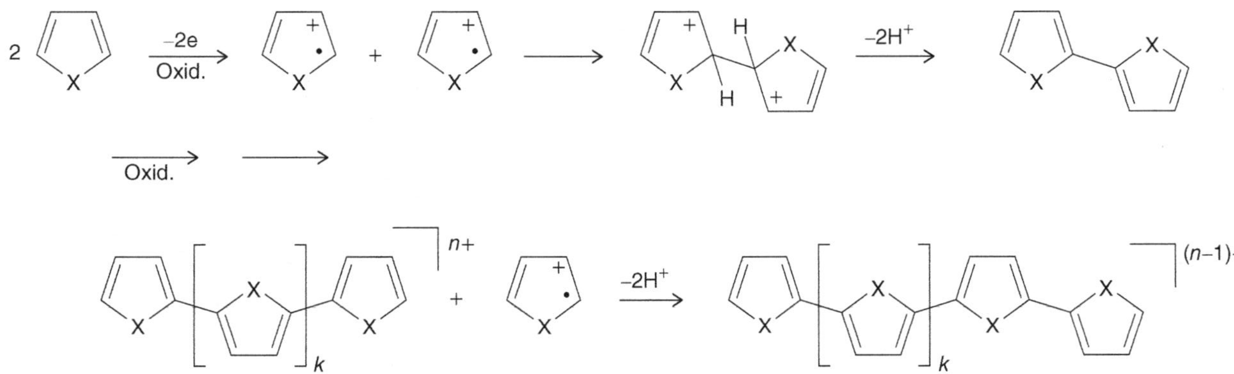

X = NH, S

Scheme 4.1 Classical formation mechanism of electrochemical polymerization for CPs. *Source:* Heinze et al. [2b]. © American Chemical Society.

deposition time, concentration, and electrode system as the key factors. These factors have a huge impact on the morphology, mechanics, and electron transport properties of polymer films. So far, in order to obtain a high-quality polymer film, much work has been devoted to the optimization of molecular structure, selection of solvents and electrolytes. To obtain films with different structures and functions, functionalized molecules usually need to be designed and synthesized. In addition, an excellent electrochemical polymerization environment is of great significance for obtaining high-quality films.

Generally, the electrochemical polymerization of aromatic compounds (such as thiophene, furan, benzene, carbazole, fluorene including their derivatives, etc.) is performed in an organic solvent system due to their insolubility or poor solubility in water. The mechanism of electrochemical polymerization confirms that the as-deposited produces at electrodes are composed of a mixture of species with various lengths of conjugated chain. Compared with normal polymerization reactions, the electropolymerization of conjugated molecules is a fully activated process, the as-deposited polymers with different chain lengths depend on experimental parameters such as ambient temperature or applied potentials [3]. Under a low oxidation potential, the as-electrodeposited PPy is an oligomer with 8–16 units in its conjugated chains. While a higher oxidation potential is applied, the chain lengths for PPy are allowed to increase to 32–64 units. Moreover, at lower polymerization temperatures, it is slower for the electrodeposition reaction kinetics leading to shorter chain lengths, however, the as-deposited polymer films are allowed to have more time to generate ordered structures, at least on a short range. The generation of order probably has a relevant impact on the transport properties of as-deposited polymer films.

Apart from the applied potential, the oxidation potential of an aromatic compound also depends on its structure and the nature of the electrolyte during electrochemical polymerization [4]. Previous studies have shown that the level of monomer oxidation potential has a greater impact on polymer performance, because the decrease in monomer oxidation potential is positively related to the conductivity of the polymer obtained finally, and causes the red shift of the polymer's maximum absorption peak indicating the number of effective conjugated links increases. At the same time, the reduction of the oxidation potential affects the stability of the conductive state of the polymer. The reduction of the monomer oxidation potential also allows many more active metals to be used as electrode materials, thereby reducing the cost of material synthesis. Therefore, a very important research content in the electrochemical synthesis of many conjugated polymers is how to reduce the oxidation potential of the monomer. There are two ways to achieve this goal: (i) Change the structure of the monomer, that is, β-position substitution with alkyl, alkoxy, halogen, and other functional groups. The thiophene monomer with a strong electron-withdrawing group has a high oxidation potential, which requires a high polymerization potential, many side reactions, and low polymerization current efficiency. Substitution of the electron-donating group reduces the oxidation potential of the monomer and is easy to stabilize the intermediate free radical cation. (ii) Change the synthesis conditions including solvents, anode materials, electrical, and magnetic conditions. Finding the right solvent and supporting electrolyte is very important to reduce the oxidation potential of the monomer and thus improve the performance of the polymer, especially for free-standing films.

Inserting a substituent group at the β-position can effectively prevent α–β linkages between monomers during the electrochemical polymer process. The introduction of electron-donating groups or denser aromatic ring molecules can better stabilize the cationic radicals, thereby effectively reducing the oxidation potential of the aromatic ring molecules. As one of the most effective methods, the functionalization of aromatic ring molecules has become the most important way to obtain high-quality polymer films [5]. When an alkyl chain (methyl or hexyl) is introduced at the β-position in thiophene, the oxidation potential of the aromatic ring molecule will be obviously reduced [6]. The introduction of these substituents is indeed a good thing for the preparation of the polymer itself, as they do reduce the oxidation potential of the polymer and stabilize its oxidized formation to avoid to nucleophilic attack. However, the substituents will severely cause steric interactions, which will lead to the deformation of the π-conjugated chains, and ultimately reduce the conductivity of as-deposited polymer in the applications of electron materials [2a]. Roncali et al. [7] found that a cyclization between 3- and 4-positions on the thiophene ring reduced the steric hindrance to conjugation in the poly(3, 4-dialkylthiophenes). Furthermore, the 3,4-alkylenedioxythiophene family were developed, especially for 3,4-ethylenedioxythiophene (PEDOT), which allows the oxidized form even more stability with higher electrical conductivity due to less steric distortion compared with other PThs [8]. A similar substitution has already been studied on pyrrole leading to the appearance of the poly(3,4-alkylenedioxypyrrole) and poly(3,4-alkylenedioxy selenophenes) families with the enhanced stability in its doped state to ambient reduction [9]. Although these CPs exhibit superior properties, a slow progress has been made owing to synthetic challenges. That is anticipated a more advanced synthesis technology is developed and availability due to the unique properties and applications. In addition, a high-quality free-standing film can also be obtained by electrodeposition based on the cross-linking molecule monomers. The free-standing PTh-based polymers with oligo(oxyethylene)-bridged 2,2′-bithiophenes as monomer (Figure 4.1a) were obtained by electrochemical polymerization in the propylene carbonate solution [5a, b, 10]. The resultant polymer films showed a high electrical conductivity and good environmental stability, which are expected as a promising flexible thermoelectric material.

For the synthesis of a high-quality polymer, solvents and electrolytes are another very important aspect. The electrosynthesis of CPs is mainly performed in organic solvents or aqueous solutions. Some aromatic compound monomers in these media cannot be polymerized due to the high oxidation potentials leading to the poor-quality polymer films. Some electrolyte molecules with electrocatalytic properties can activate monomers and effectively reduce their oxidation potentials. As we know, thiophene monomer

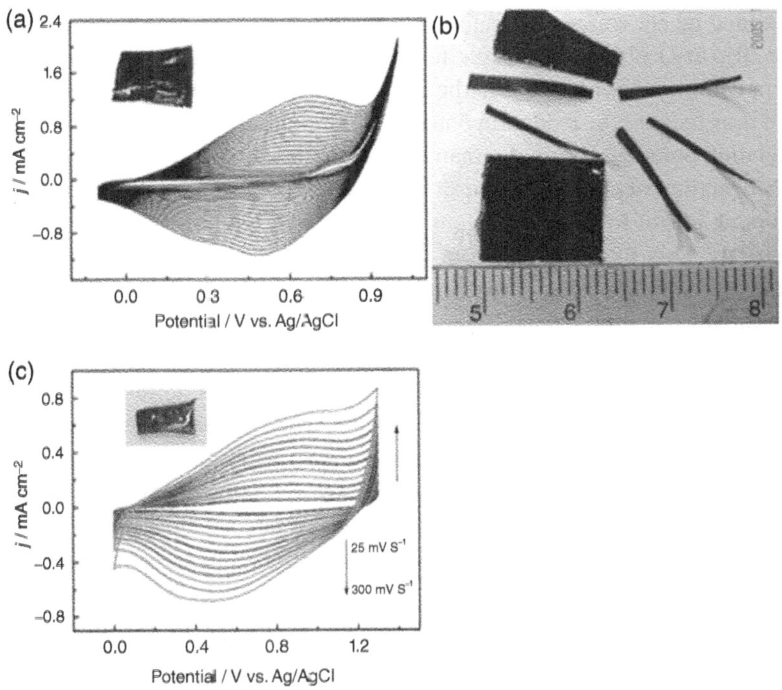

Figure 4.1 Free-standing conductive plastic obtained by electrochemistry. (a) Cyclic voltammograms (CVs) of oligo(oxyethylene)-bridged 2,2′-bithiophenes in PC/$LiClO_4$. *Source:* Hu et al. [10]. © Elsevier. (b) Poly(1,12-bis(carbazolyl)dodecane) film image. *Source:* Xu et al. [11]. © Elsevier. (c) CVs of thiophene-furan. *Source:* Yao et al. [12]. © Royal Society of Chemistry.

has a high oxidation potential in common organic solvent (acetonitrile, dichloromethane, chloroform, propylene carbonate, water, etc.) and electrolytes ($LiClO_4$, tetrabutylammonium hexafluorophosphate, tetrabutylammonium tetrafluoroborate, etc.). Boron trifluoride diethyl etherate solution ($BF_3 \cdot Et_2O$) is currently one of the most ideal mediums, which can effectively reduce the oxidation potential of a series of aromatic monomers and obtain high-quality polymer films [13]. A high-quality free-standing PTh film was electrochemically deposited in $BF_3 \cdot Et_2O$ with an applied potential of 1.3 V (vs. Ag/AgCl) resulting in an electrical conductivity of 48.7 S cm^{-1} and a tensile strength of 1200–1300 kg cm^{-2} which was greater than that of aluminum. Although thiophene oligomers can be used as monomers in conventional solvents to reduce the oxidation potential, the conjugated chains of PThs obtained from oligomers are relatively short. Compared with common solvents, the average conjugated chain length of PTh obtained by direct oxidation of thiophene in $BF_3 \cdot Et_2O$ solution is 80 units, which is much larger than the average conjugated chain length of the product obtained by electrolysis in acetonitrile solution (~40 units). In this system, the $BF_3 \cdot Et_2O$ plays a main role to interact with thiophene monomers to form complexes, which lowers the resonance energy of the thiophene ring and thus reduces the oxidation potential of thiophene. The reduction of the oxidation potential eliminates many side reactions such as α–β linkages and the formation of carbonyl groups. At the same time, $BF_3 \cdot Et_2O$ reduces the coupling rate of free radical cations to each other, enabling it to polymerize regularly. This $BF_3 \cdot Et_2O$ medium was also used to polymerize other aromatic compounds, such as pyrrole, furan, benzene, carbazole, and their derivatives.

In order to further improve the quality of electrodeposited polymers, a mixed system of $BF_3 \cdot Et_2O$ with an alcohol, diethyl ether, tetrahydrofuran, or strong acid such as concentrated sulfuric acid, or trifluoroacetic acid is also generally used for electrochemical polymerization of monomers. For example, pyrrole and furan, because of their relatively low resonance energy, are very unstable in $BF_3 \cdot Et_2O$, and are prone to additional polymerization reactions to form nonconjugated polymers. Adding the same amount of diethyl ether to $BF_3 \cdot Et_2O$ solution can effectively reduce its Lewis acidity, so that pyrrole can smoothly undergo α–α polymerization in this system resulting in the high-quality PPy film. $BF_3 \cdot Et_2O$ can be mixed with isopropyl alcohol to form a proton electrolyte. The main conductive ion is $ROBF^{3-}$, which is a complex of $ROBF_3H$ formed by the alcohol and BF_3, and ROH^{2+}, the conjugate acid of alcohol [14]. Pyrrole in this system can obtain high-quality film by electrochemical polymerization, and its performance (~10^2 S cm^{-1}, 40–60 MPa) is better than that of other systems (10^{-2}–100 S cm^{-1}, fragile film), similar to the performance of PPy film prepared in aqueous solution or acetonitrile solution. Figure 4.1b shows that a high-strength free-standing poly(1,12-bis(carbazolyl)dodecane) (P2Cz-D) film was successfully electrodeposited in the mixed $BF_3 \cdot Et_2O$ and $CHCl_3$ (40 *vol.%*) due to internal cross-linked dicarbazole units and showed the electrical conductivity of 1.36 $S\,cm^{-1}$ and a tensile strength of 230 $kg\,cm^{-2}$ [11, 15]. In contrast, in addition to the addition of an equivalent amount of ether, the electrochemical polymerization of furan also requires the addition of an appropriate amount of BF_4^- anion.

This does not mean that it is difficult to obtain a high-quality film in a neutral solvent, but that it mainly depends on the molecular structural characteristics. Recently, Yao et al. [12] obtained a free-standing poly(thiophene-furan) (PTFu) film by the electrochemical polymerization in a neutral CH_2Cl_2 with tetrabutylammonium hexafluorophosphate (Bu_4NPF_6) as electrolyte. As shown in Figure 4.1c, the PTFu films with excellent electrochemical activity and environmental stability with an electrical conductivity of $5.5 \times 10^{-2}\,S\,cm^{-1}$ and large Seebeck coefficient of 103.5 $\mu V\,K^{-1}$. In addition, in order to obtain high-quality polymer films, some chemical reagents such as alcohols and surfactants, can also be added to the electrolyte solution. Ma et al. [16] developed mixed $BF_3 \cdot Et_2O$ and isopropyl alcohol with pentaerythritol ethoxylate (PEE) to achieve a strong and flexible PPy film which allows contractile stress up to 27 megapascals and lift objects 380 times heavier than itself. These facts demonstrate that the electrochemical deposition for the preparation of high-quality free-standing polymer films is a powerful tool with respect to the inherent advantages in the field of flexible electronic devices. Although organic molecules are rich in structure, for organic thermoelectric materials, researchers may need to consider the cost, stability, and electron transport performance of the corresponding polymer from the beginning, so as to better guide the synthesis of molecules and ultimately achieve high performance of CPs films.

4.3 Electronic Structure and Optical Properties

It is well known that electrons in metals have a high density of electronic states, which makes their binding energy relatively low. Once an electric field is faked, "free electrons" can easily move among atoms, thereby achieving a high electrical conductivity. A key characteristic for CPs is the presence of a strong chemical single σ-bond and a less strongly localized double π-bond alternations between carbon atoms with sp^2 hybrid along the main chain. For polyacetylene (PAc) as an example, there are three electrons in each carbon atom outer shell to form three σ-bonds and one 2p electron to form a π-bond. The 2p electrons produce delocalization in π-orbitals with an electronic density dominant in the planar σ-skeleton, where its stabilization allows a large torsion barrier [17]. CPs as semiconductor have a more rigid chain and a smaller band gap compared to that of saturated polymers.

In CPs, the formation of π and π^* antibonding states from the p atomic orbitals is predicted after the formation of molecular orbital based on Hückel's molecular orbital theory. The highest occupied molecular orbital (HOMO) is defined as valence band (VB), the lowest unoccupied molecular orbital (LUMO) is denoted as conduction band (CB). The structure of four molecules is showed with different conjugated chains from ethylene to polyacetylene [18]. The ethylene has a large HOMO–LUMO gap. As the increase in the length of conjugated chains (long enough), energy bands (1–4 eV) with convergence are created and becomes smaller when molecular orbitals are infinitely close resulting in the semiconducting polyacetylene. Generally, the longer the conjugated chain, the more the states created in these, resulting in smaller energy gap between HOMO and LUMO. PTh also shows the same change trend that their band gap will become smaller with the increase of thiophene rings in the conjugated chain. In addition, the structural modification has a significant impact on the electronic properties of CPs. For PTh, the experiment and the calculated density of states (DOSs) demonstrated the fact that the electron-withdrawing group can lower the HOMO–LUMO gap. It is worth noting that the electronic structure of CPs is capable of being affected by steric repulsion between substituents, which also can be lowered by introducing a vinyl bridge in conjugated chain between thiophene rings [19].

According to the quantum-mechanical model, the energy for HOMO can be defined as [20]:

$$E(\text{HOMO}) = (N/2)^2 h^2/8m(Nd)^2$$

where, h is the Planck's constant, m is the electron mass, N is atom number, and d is the distance between two atoms. The LUMO has the energy:

$$E(\text{LUMO}) = (N/2+1)^2 h^2/8m(Nd)^2$$

The energy required to excite from HOMO to LUMO (ΔE) is thus:

$$\Delta E = E(\text{LUMO}) - E(\text{HOMO}) = (N+1)h^2/8m(Nd)^2 \approx \left[h^2/8m(Nd)^2\right]/N$$

If a photon satisfies this excitation energy, its energy can be expressed as:

$$\Delta E = E(\text{LUMO}) - E(\text{HOMO}) = h\nu = hc/\lambda$$

where λ is the wavelength, ν is the frequency of light, and c is the velocity of light.

The optical absorption of undoped polyacetylene has a small band gap of 1.7 eV indicating a classical semiconductor. Although the theory has been successful in predicting optical behavior, there are also experiments suggesting that the model is not universally valid as the hypothesis is conceived for a one-dimensional linear structure without defects. The actual situation is that the electrical conduction in CPs not only exists linearly, but also includes the conduction by inter-chains which can be explained through the hopping process. Moreover, only to rely on conjugation is not enough to achieve a high electron transport capacity in the intra-chains of CPs. It is another very important aspect in improving the transport property of electrons by the weak covalent π–π interactions from inter-chains and oxidation levels.

A typical CP is allowed to have the different oxidation states and the aggregation between conjugated chains leading to the broad, shifted, various intensities absorption features. The electrons, until doping, remove from the VB of neutral chain and jump to the bottom of the CB, which results in the formation of polarons (radical cation) attributed to the two new in-gap states. After further removal of the electrons from the top of VB, the new bipolarons (dication) appear in a conjugated chain. The new optical transition level for a polaron consisting a 1/2 spin is half-filled orbitals, while it is empty for a bipolaron without spin. It is worth noting that the formed bipolarons are not simply composed of two polarons. The charged defects are balanced through the negative counterions introduced during the synthesis. However, for the aggregation of chains in the solid polymer, it is not easy to fully determine the transitions of polarons and bipolarons. With the increase in oxidation level, the (bi)polaron levels of in-gap states are spatially localized on a chain or assembly of chains (intrachain or interchain) for a polymer in an amorphous phase, which leads to that the Fermi level (E_F) lies in the middle of the polaron band or between the VB and the bipolaron band for a disordered polaronic and bipolaronic polymer, respectively [21]. The electronic structure in disordered (bi)polaron polymer solid can be considered as Fermi glasses. For crystalline domains of CPs, short interchain distances will cause an overlap of π-electron density, thereby promoting the delocalization of the electron wave function and over several chains and resulting in more metallic behavior. Regarding this part of knowledge, Bubnova et al. [17, 21] did a very detailed discussion and explanation for CPs.

At present, the technologies that can theoretically obtain information about the energy level structure and electronic properties of semiconductor materials include: space charge spectroscopy, X-ray photoelectron spectroscopy (XPS), ultraviolet photoelectron spectroscopy (UPS), scanning tunneling microscopy (STM), theoretical calculation, cyclic voltammogram (CV), spectroelectro-chemical technique, etc. According to Planck equation, the onset values at the lowest energy bands (λ_{onset}) is related to the optical $E_{g,opt}$:

$$E_{g,opt} = \frac{1240}{\lambda_{onset}} = E_{LUMO} - E_{HOMO}(\mathrm{eV})$$

In comparison, the CV method can not only characterize the dispersed state of the solution, but also study the electrical properties of the film-forming aggregates. At the same time, it has obvious advantages in terms of cost and operability, so it is particularly suitable for characterizing the energy level structure of nanomaterials and electronic characteristics. In fact, the CV method has been widely used to characterize the energy level structure of conjugated polymers [22]. Figure 4.2 shows the measurement principle. In the electrochemical CV test, the oxidation process of CPs corresponds to the process by which the electrode draws electrons from its VB. The onset of oxidation potential (E_{ox}) relative to reference electrode corresponds to the HOMO energy level (the top of VB, ionization potential I_p). During the process of injecting electrons in the medium, its onset of reduction potential (E_{red}) corresponds to the CB bottom or LUMO energy level. They have the following correlation equations:

$$E_{HOMO} = -e(E_{ox} + V_{ref})(\mathrm{eV})$$

$$E_{LUMO} = -e(E_{red} + V_{ref})(\mathrm{eV})$$

where V_{ref} is the distance of the electrode potential of the reference electrode relative to the vacuum level. If ferrocene (Fc/Fc^+) is used as a reference, the value of V_{ref} is 4.8 eV; if saturated calomel electrode (SCE) is used as a reference, V_{ref} is 4.4 eV. In addition, since the polymer's energy level structure and electronic properties are affected by multiple factors such as its size, morphology, and structure, in theory, the CV test covers a wealth of information on microstructure and electronic properties.

Both theoretical and experimental studies have demonstrated that the increase of polarons and bipolarons introduced into polymer by donor or

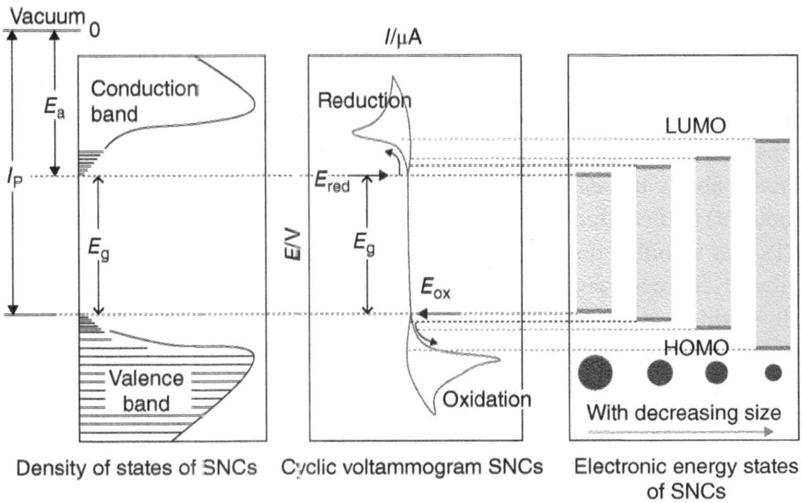

Figure 4.2 Illustration of the principle of measurement for the HOMO and LUMO energy levels by electrochemical CV method. *Source:* Li et al. [23]. © Progress in Chemistry.

acceptor doping, can promote the large enhancement of electrical and thermal conductivity of CPs. This means that polarons and bipolarons can be created and regulated through doping, suppressing, or adding charge to optimize thermoelectric transport property. Therefore, CPs are allowed to achieve a metallic electrical conductivity, as a promising thermoelectric material showing higher Seebeck coefficients and lower thermal conductivity than metals. During recent years, in order to clarify the relationship between electrical transport and electronic structure and morphology, many works have been carried out especially for thiophene-based polymers. Because the DOS is directly related to the electrical conductivity and its slope at E_F is related to the Seebeck coefficient, the understanding of DOS can better understand the thermoelectric transport property. It is worth noting that, despite the fundamental importance of this issue, few theories are used to study the microscopic research work on DOS in realistic CPs owing to the requirement of multiscale modeling. Although Muñoz et al. [24] speculated about the evolution of the band structure in semi-metallic PEDOT, indicating that the bipolaron band can be incorporated into the delocalized VB, there is no theory to explain this remarkable behavior so far. New models have yet to be developed to try to overcome the shortcomings of classical polaron and bipolar models.

4.4 Electrochemical Doping and De-doping

CPs with the presence of conjugated structure consisted of an alternating single-double bond resulting in the semiconducting properties. One of the most important characteristics of CPs used in thermoelectric materials is the capability to easily realize the reversible conversion between undoped (neutral) and conducting or doped (oxidized) states. Conjugate systems in CPs can stabilize positive/negative charge carriers by delocalization. At present, there are fewer n-doped CPs, mainly due to their poor stability in air, so this section does not discuss the n-doping. In contrast, most of the current research focuses on the p-doped CPs, which is obtained through the introduction of anions, and has relatively good stability in room temperature environments. Therefore, this also means that CPs can change its doping levels and types by physical, chemical, and electrochemical methods. There have been many reports of physical and chemical methods of doping, which will not be described here. The focus is mainly on the electrochemical doping for CPs.

The electrochemical method has unique advantages in the redox control of CPs, because electrochemical stoichiometry can be easily manipulated by electrochemical instruments. The electrochemical redox of CP is usually carried out in an electrolyte solution system with the movement of ions into and out of the polymer matrix. It is worth noting that the electrochemical doping efficiency is related to the thickness of the CP film. Generally, as the thickness of the CP film increases, the doping speed becomes slower, which is mainly due to the limited ion transport in the CPs [25]. Its oxidation mainly manifests itself as doping of anions to balance electron transport, while reduction mainly involves detachment of anions from the chain in Figure 4.3. By applying an application potential, the existence and detachment of anions in CP can be effectively realized, and the type of electrolyte can be easily changed to change the corresponding anion dopant, thereby regulating the electron transport performance of CPs. In the neutral state, CPs exhibit very low electrical conductivity (10^{-9}–$10^{-4}\,\mathrm{S\,cm^{-1}}$), which can be greatly improved through the increase of the doping level. For a good thermoelectric CP, it is necessary to achieve as high an electrical conductivity (σ) as possible with regard to a large Seebeck coefficient (S). In semiconductor materials, because conductivity is usually directly proportional to carrier concentration, Seebeck coefficient is inversely proportional to it. Compared with inorganic thermoelectric materials, CPs usually have the lower thermal conductivity ($< 1.0\,\mathrm{W\,m^{-1}\,K^{-1}}$). Although the doping level has a certain effect on its thermal conductivity, the thermoelectric power factor (σS^2) is usually used to evaluate its thermoelectric performance.

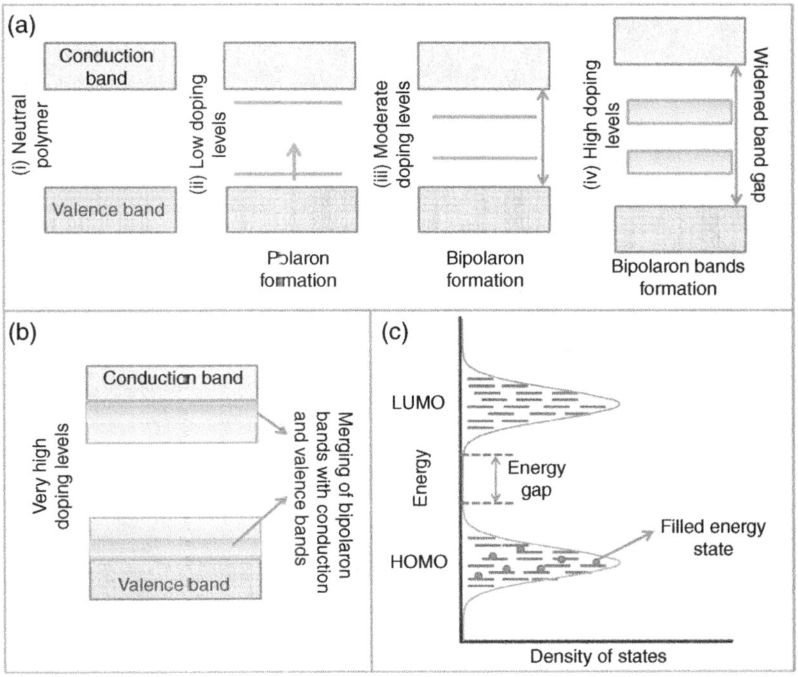

Figure 4.3 Schematic showing (a) Formation of polaron, bipolaron, and bipolaron bands as a function of doping for CPs; (b) Merging of bipolaron bands with conduction and VBs at very high doping levels; (c) Gaussian distribution of localized states in HOMO and LUMO orbitals of CPs. *Source:* Bharti et al. [26]. © Elsevier.

Therefore, effective optimization of doping level is one of the main ways to improve the thermoelectric power factor of CPs.

Scheme 4.2 reveals that the doping in PEDOT can induce new electronic states (polarons or bipolarons) in the bandgap of CPs, which is achieved by a redox reaction related to electron transfer. A high HOMO energy level directly related to the optical properties can provide a low oxidation potential with the environmental stability for CPs. As shown in Figure 4.3, the p-doping involves an electron transfer from the HOMO to the LUMO by an electrochemical redox. With the respect to a particular dopant, molecular engineering is also often used to optimize HOMO and LUMO levels. The band level in CPs can be adjusted to facilitate π-electron flowing along the polymer backbone by the design of donor-acceptor copolymerization, the introduction of electron-withdrawing groups, and fused heterocycles to improve backbone planarity and extend the delocalization as well as a two-dimensional conjugation [27]. For thermoelectric CPs, an effective p-doping requires reasonable design or manipulation of its HOMO and LUMO levels

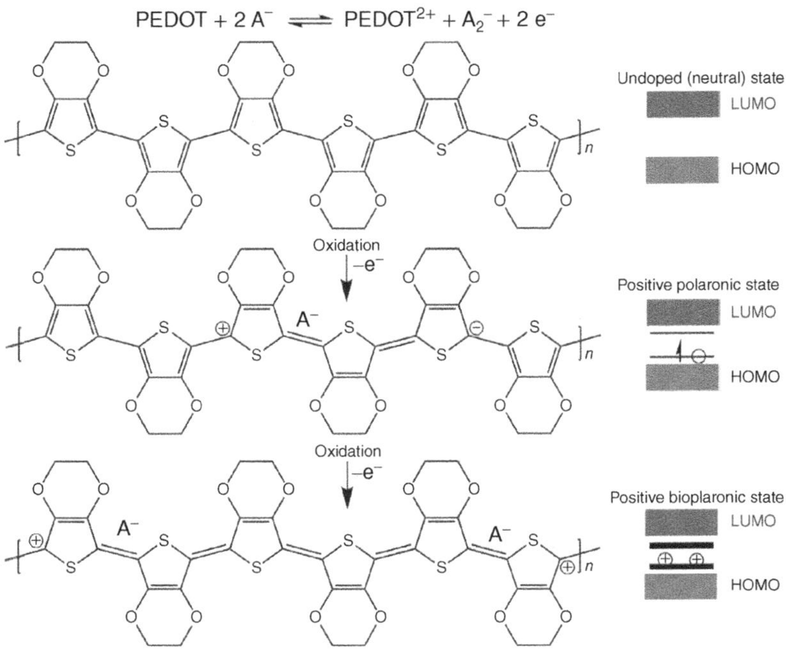

Scheme 4.2 Illustration for p-doping in PEDOT with their chemical and band structure evolution. The oxidation ($-e^-$) from undoped (top) to doped states results in the formation of positive polaronic (middle) and bipolaronic (bottom) states. Anion counterions are denoted as A^- in the CP chains.

to achieve a slight offset. For example, a highly doped CP has a narrow semi-metallic band, although higher electrical conductivity may be obtained, their Seebeck coefficient is usually very small, which has been experimentally confirmed in PAc, PANi, PTh, PPy including their derivatives. In contrast, a CP with a low doping level has a larger Seebeck coefficient depending on the difference in energy between Fermi level and the transport level, even up to several hundreds of $mV\,K^{-1}$ [17]. It is distinctly different for the localized polaronic states in ordered (an energy distribution within kT) and disordered (an energy range larger than kT) CPs. An excellent thermoelectric CP must require a highly stable doping state, a reasonable doping level, and an effective doping anion.

In addition, the doping level (f) of the polymer film can be determined by electrochemical method, and the calculation formula is defined as:

$$f = \frac{2Q_0}{Q_d - Q_0} \times 100\%$$

where Q_d is the total charge for the deposition and Q_0 is the total charge of oxidized species in as-prepared CPs films obtained by electrochemistry.

Xu et al. [6a] studied the relationship between the doping degree of PTh and its derivative films with film thickness by electrochemical methods. Experimental results indicate that in the early stage of polymer growth, whether it is poly(3-chlorothiophene) or poly(3-methylthiophene), their doping degree is close to zero, that is, it is in an approximately insulating state. After that, their doping degree increases slowly with the increase of the film thickness, and their doping degree does not stabilize and reach a constant value until the thickness exceeds a certain value. In addition, the maximum doping degree of the polymer film in the BF_3•Et_2O solution was 28% poly(3-methylthiophene) and 19% poly(3-chlorothiophene). At the same time, the film thickness of poly(3-methylthiophene) required to reach the maximum doping level is about 500 nm, while poly(3-chlorothiophene) is about 2000 nm. The results show that substitution of the methyl group to the electron group can increase the doping degree of PTh, while substitution of the chlorine atom of the electron withdrawing group is not conducive to the high doping degree of PTh. The main reason that the doping degree of the PTh derivative film increases with the increase in thickness may be due to the morphological change during the growth process.

4.5 Thermoelectric Performance of Flexible CP Films

4.5.1 Polythiophenes

The thermoelectric CPs have been extensively investigated due to their low cost, low thermal conductivity [28], and rich structure [29]. More importantly, CPs have a widely adjustable doping level, which provides a vast research space for the optimization of thermoelectric performance. The electrochemical method not only has unique advantages in the preparation of high-quality CPs, but also can effectively regulate its doping level through electrochemical stoichiometry control to tune their thermoelectric performance. There are many factors influencing the thermoelectric properties of CPs. In order to avoid the influence of structural changes as much as possible, it is desirable to obtain the corresponding polymer by electrodeposition and then doping or de-doping it in the corresponding solution.

Thiophene-based CPs is the most typical representative in the field of organic thermoelectric materials. Experimental results have demonstrated that a high-quality PTh obtained by electrodeposition can be regarded as a promising thermoelectric CP, as shown in Figure 4.4 [28b]. A free-standing PTh film from BF_3•Et_2O solution showed the high electrical conductivity of $47\,S\,cm^{-1}$ (Figure 4.4a) and the large Seebeck coefficient of $130\,\mu V\,K^{-1}$ (Figure 4.4b) as well as a low thermal conductivity of $0.17\,W\,m^{-1}\,K^{-1}$ (Figure 4.4c) with the high tensile strength, which is unmatched by chemical preparation methods. Its thermoelectric figure-of-merit (*ZT*) was up to 3.0×10^{-2} at 250 K (Figure 4.4d) and higher compared to that of many other CPs. Although the electrical conductivity of the PTh thin film gradually decreases with time delay, its *S* significantly increases, and eventually its *ZT* value can be maintained in the order of 10^{-2} after two months. The experiment also verified that the thermoelectric *ZT* value of the free-standing poly (3-methylthiophene) (PMeT) film is similar to that of PTh, both of them exhibit better thermoelectric performance than their pellets obtained from powder. A free-standing flexible poly(3,4-ethylenedioxythiophene) (PEDOT) derivation (PBEDTH) was electrodeposited in CH_2Cl_2 containing 0.1 M Bu_4NBF_4, which shows a higher Seebeck coefficient of $37.2\,\mu V\,K^{-1}$ and lower electrical conductivity of $0.11\,S\,cm^{-1}$ compared to PEDOT at 300 K [30].

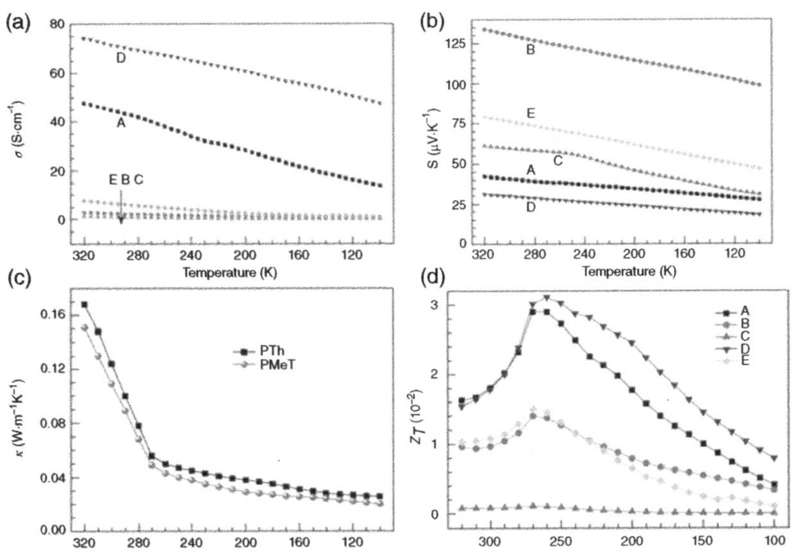

Figure 4.4 Temperature dependence of the electrical conductivity (a), the Seebeck coefficient (b), the thermal conductivity (c), and the *ZT* values (d) of PTh and PMeT films by electrochemical deposition in BF_3•Et_2O. The fresh free-standing PTh film is dried for 24 hours denoted as sample A and placed for 2 months in air as sample B. The PTh pellets are obtained from powder under 20 MPa assigned as sample C and D after two months. Sample E is the free-standing PMeT film after two months. *Source:* Lu et al. [28b]. © Chinese Physical Society.

These facts suggest that the electrochemical deposition provides a method for the preparation of CP thin films with high thermoelectric performance, at least this free-standing film is powerful for practical application in electron devices.

The microstructure of CP films is affected by the molecular structure and electrolyte, which is also related to their electron transport performance. Thieno[3,2-*b*]thiophene (TT) composed of two condensed thiophene rings, is a very common organic molecule in organic optoelectronic materials, which is often used to improve the planar rigidity of polymers to optimize carrier mobility. For a CP, the introduction of TT units in the conjugated chains allows a better ordered polymer structure and an enhanced charge carrier mobility due to good π-π packing, resulting in the simultaneous increase in electrical conductivity and the Seebeck [31]. The chemical oxidation for TT generally gains the oligomers. A free-standing polythieno[3,2-*b*]thiophene (PTT) has been prepared by electrochemical deposition in $BF_3 \cdot Et_2O$ solution with a thermoelectric power factor of $1.1\,\mu W\,m^{-1}\,K^{-2}$ at room temperature [32]. One can find that the electrolyte solution system has a greater effect on the morphology of PTT. The introduction of different alkyl chains at the β-position of thiophene greatly changed its electrical conductivity and Seebeck coefficient. Compared to PTh, PMeT exhibits a higher electrical conductivity ($95.8\,S\,cm^{-1}$) and a smaller Seebeck coefficient ($14.9\,\mu V\,K^{-1}$), while the poly(3-octylthiophene) has the smaller values in both electrical conductivity and Seebeck coefficient [33]. In CP semiconductors, the relationship between electrical conductivity and Seebeck coefficient is very delicate, and it is difficult to predict its thermoelectric performance simply based on its chemical structure.

4.5.2 Polyselenophenes

Selenophene has the analogue of thiophene with the replacement of the sulfur atom by the more polarizable selenium atom and a lower oxidation potential for the corresponding polymer. Owing to the unique properties of the Se atom and selenophene, polyselenophene (PSe) may have many advantages over PTh such as a lower band gap, more quinoid character, harder to twist [9d]. Nevertheless, experimental results show that PSe prepared by either electrochemical or chemical oxidation has a very small electrical conductivity ($\sim 10^{-2}\,S\,cm^{-1}$), which may be attributed to the poor stability during the oxidative polymerization of selenophene monomers and the lack of suitable precursors [9d]. For electrodeposition, the quality and characteristics of PSe films strongly depend on the solvent-electrolyte combination. However, the high-quality free-standing PSe film has been obtained by electrochemical polymerization in $BF_3 \bullet Et_2O$ and ionic liquid with the electrical conductivity of 0.28 and $2.3\,S\,cm^{-1}$, respectively. Interestingly, PSe film shows a high Seebeck coefficient ($>180\,\mu V\,K^{-1}$) with a low electrical conductivity of $10^{-2}\,S\,cm^{-1}$ [34]. Until recently, Kim et al. [35] developed the poly(hexyl-3,4-ethylenedioxyselenophene) (PEDOS-C6) by electrochemical polymerization at a constant potential of 1.1 V in PC solution containing 10 mM selenophene monomers and 0.1 M $LiBF_4$. The as-deposited PSe film shows the high electrical conductivity ($180\,S\,cm^{-1}$), Seebeck coefficient ($30\,\mu V\,K^{-1}$), and power factor ($16\,\mu W\,m^{-1}\,K^{-2}$) generated from the vertical direction. This result is comparable to that of PEDOT obtained by chemical vapor-phase polymerization [36]. It can be believed that the charge transfer in PSe is more suitable to occur between inter-chains by intermolecular Se···Se contacts.

4.5.3 Polycarbazolyls

The carbazole-based polymer exhibits good photoconductivity and stable hole-transport properties, so it has attracted much attention in the research of optoelectronic materials [37]. Poly(2,7-carbazole) has a lower HOMO energy level (−5.95 eV) estimated according to its oxidation onset than that of the air oxidation threshold −5.27 eV (0.57 V vs. SCE), indicating the good air stability. Theoretical calculation and experimental results also have suggested that poly(2,7-carbazole) is a promising organic thermoelectric material. This type of polymers including N-substituted derivatives can easily obtain a high-quality self-supporting film by electrochemical deposition [11, 15, 38]. Moreover, N-substituted poly(2,7-carbazole)s allow to have the oxidation in nitrogen atom occurring priority to conjugated backbone, which results in a possibly large Seebeck coefficient and a small electrical conductivity due to charge carrier pinning.

Xu et al. [11] successfully obtained a homogeneous and high-strength free-standing poly(1,12-bis(carbazolyl)dodecane) film. It is observed that the as-prepared film gains the electrical conductivity of $2.65 \times 10^{-6}\,S\,cm^{-1}$, which is much smaller than the pre-exponential factor ($\sigma_0 \sim 10^4\,S\,cm^{-1}$), revealing the conduction in the polymer chains by hopping in the localized states [15]. The activation energy for electrical conductivity (0.25 eV) is much higher than that in the thermopower, further confirming that the charge carriers in poly(1,12-bis(carbazolyl)dodecane) move by hopping in the localized states at band gap edges. Theoretical calculations suggest that a thermoelectric material with good performance requires a high Seebeck coefficient related to a flat band near the Fermi level and a large electrical conductivity depending on a dispersive band. Although the opposite relationship between electrical conductivity and Seebeck coefficient can be resolved by introducing some groups (such as vinylidene, thiophene, EDOT, etc.) on the side chain and backbone, this is outside the scope of this discussion. Compared to chemical method, electrochemical methods may require more effort on the ordered structure of polymer molecules in solid state.

4.5.4 Copolymers

Electrochemical copolymerization is an effective method to optimize the molecular structure and electron transfer properties of polymers. It has been widely used in the preparation of organic optoelectronic and thermoelectric materials. This method can connect different structure molecules to the conjugated backbone in order to improve the related properties by changing the ratio of different molecules in solution. Yao et al. [12] designed and synthesized the thiophene-furan monomer to prepare a free-standing polymer (PTFu) by electrodeposition in neutral CH_2Cl_2-Bu_4NPF_6 electrolyte system. The resultant pristine PTFu film shows the higher Seebeck coefficient (74.1 $\mu V K^{-1}$) and electrical conductivity (1.2×10^{-3} $S cm^{-1}$) than PTh and polyfuran (PFu) in a neutral medium. This copolymerized structure of PTFu probably generate a flatter band near the Fermi level compared to PTh and PPy.

From the perspective of molecular structure, various precursors can be selected for copolymerization to obtain thermoelectric copolymer materials. For example, for the electron donor/acceptor alternating copolymers with narrow bandgaps, poly(4,7-bis(2,3-dihydrothieno[3,4-b][1,4]dioxin-5-yl) benzo[c][1,2,5]thiadiazole) (PEBTE) and poly(4,7-bis(2,3-dihydrothieno[3,4-b] [1,4]dioxin-5-yl)benzo[c][1,2,5]selenadiazole) (PEBSeE) had fair electrical conductivity (10^{-1}–10^{1} $S cm^{-1}$) and Seebeck coefficient (14.0 $\mu V K^{-1}$) [39]. Taking into account the planar of molecular chains, poly(1,12-bis(carbazolyl) dodecane-co-thieno[3,2-b]thiophene) (P(2Cz-D-co-TT)) showed the enhanced Seebeck coefficient and power factor than its homopolymers [40]. In terms of molecular polarity, the as-deposited copolymers consisting of thiophenes and selenophene exhibited a remarkable increase in the electrical conductivity (10^{0}–10^{1} $S cm^{-1}$) with acceptable Seebeck coefficient (45–98 $\mu V K^{-1}$), resulting in significantly improved ZT value as high as 3.4×10^{-2} [34]. 3,4-Ethylenedioxythiophene (EDOT) as a popular precursor is often used to combine with thiophenes [5b, 6c], selenophene [41], and 1,12-bis(carbazolyl) dodecane (2Cz-D) [42], and both of them show the improved thermoelectric performance. Therefore, the copolymerization method can obtain a polymer with better thermoelectric performance by optimizing the molecular structure, which is a simple, efficient, and desirable method.

4.6 Control in Thermoelectric Performance by Electrochemistry

Besides electrochemical synthesis, the electrochemical control in the doping level is a very useful way to optimize the thermoelectric performance of CPs. We introduced the electrochemical deposition of elastic plastics in aforementioned discussion. Here, the discussion focuses on the electrochemical control in thermoelectric performance. Compared with chemical method, electrochemistry typically allows better control of the oxidation level in CPs through tuning of applied potentials. Moreover, various counterions to balance the doping charges are available simply with different salts as electrolyte [43]. The electrochemical regulation has not attracted widespread attention attributed to: (i) the shape of the regulated object is high, such as self-supporting, and the material needs to be able to effectively adhere to the electrode surface or serve as an electrode; (ii) The material is required to have good environmental stability, since electrochemistry is mainly performed in a solution system; (iii) The material itself has good electrochemical activity; (iv) It is difficult to manipulate changes in its microstructure, such as the orderly arrangement or accumulation of molecules.

In general, CPs deposited by electrochemical anodization is in a deeply doped state and exhibit good conductivity, depending on the high carrier concentration. As we know, the thermoelectric performance of a good material is not only directly related to the electrical conductivity, but also positively related to the square of Seebeck coefficient. However, the electrical conductivity and the Seebeck coefficient are inversely related to the charge carrier concentration, which means that the high-performance thermoelectric materials should have a moderate carrier concentration. Electrochemistry has been proven by many experiments to be an effective way to regulate the doping level of CPs. That the thermoelectric performance of free-standing poly(3-methylthiophene) (P3MeT) film significantly depended on its doping level [6b]. An optimum electrical conductivity ($78.9\,S\,cm^{-1}$) and a significant increase in the Seebeck coefficient ($64.3\,\mu V\,K^{-1}$) have been achieved at 0.5 and -0.5 V, respectively. Kim and his co-worker have done excellent works on the control in the thermoelectric performance of CPs film by electrochemistry [35, 44]. The pristine PEDOS-C6 film obtained by electrodeposition gained a low power factor of

$16\,\mu W\,m^{-1}\,K^{-2}$, in the vertical direction [35]. When a constant potential was applied during a certain range, the electrical conductivity increased sharply initially, then reached relative saturation, while Seebeck coefficient showed exactly the opposite change. This optimization of power factor also has been achieved for the PP-PEDOT from 862.9 to $1270\,\mu W\,m^{-1}\,K^{-2}$ at 0.1 V [44].

The thermoelectric performance of copolymer PTFu was investigated by electrochemistry (Figure 4.5a and b) with electron paramagnetic resonance (EPR) [12]. EPR is a powerful tool to determine the polaron and bipolaron changes and carrier nature in polymers. As the increase in oxidation potential, EPR signal exhibits rapid increase and reaches the maximum at 0.4 V, demonstrating the effective control in the doping level by electrochemistry. Noted that many efforts have revealed that a higher electrical conductivity in CPs usually occur with a higher concentration polaron compared to that of bipolaron. Although, PTFu film shows a low power factor of $6.6\times10^{-4}\,\mu W\,m^{-1}\,K^{-2}$, the optimized thermoelectric performance has been achieved by electrochemical doping. Therefore, the electrochemical (de) doping is sensitive to an applied potential allowed as a powerful tool to tune thermoelectric performance.

In addition, it is well known that the good order structure also plays an important role in the improvement of carrier mobility directly related to the thermoelectric performance for CPs. Generally, the CPs obtained by electrochemical polymerization exhibit amorphous structure. Santoso et al. [45] synthesized the layer-by-layer structure PTh film (Figure 4.5c) in $BF_3\bullet Et_2O$ solution without and with anionic surfactant of sodium dodecyl sulfate (SDS) and sodium dodecyl benzene sulfonate (SDBS). The introduction of surfactants with the anion tail enables PTh chains to create steric effects and reduce the packing density. However, the achieved increase in electrical conductivity ($64.9\pm3.9\,S\,cm^{-1}$) is ascribed to the dual role of surfactants as dopants and templates for molecular ordering (Figure 4.5d). Unfortunately, few efforts have been focused on the microstructural properties in order assemble by electrochemical deposition. Further it is hoped that more attention can be attracted in the investigation of thermoelectric performance for the ordered CPs by electrochemistry.

Alternative approach to the electrical conductivity and Seebeck coefficient is modifying the redox state in CPs by organic electrochemical transistors (OECTs). This is because the actual electronic transport for OECT not only occurs in the bulk, but also along an interface. Many works have illustrated the mechanism of OECT elsewhere. Bubnova et al. [43] reported an OECT both to control the doping level and measure its thermoelectrical performance of PEDOT in situ. As the increase of gate voltage (V_G), the electrical conductivity gradually reduces exponentially ($224\,S\,cm^{-1}$ at $V_G = 0\,V$),

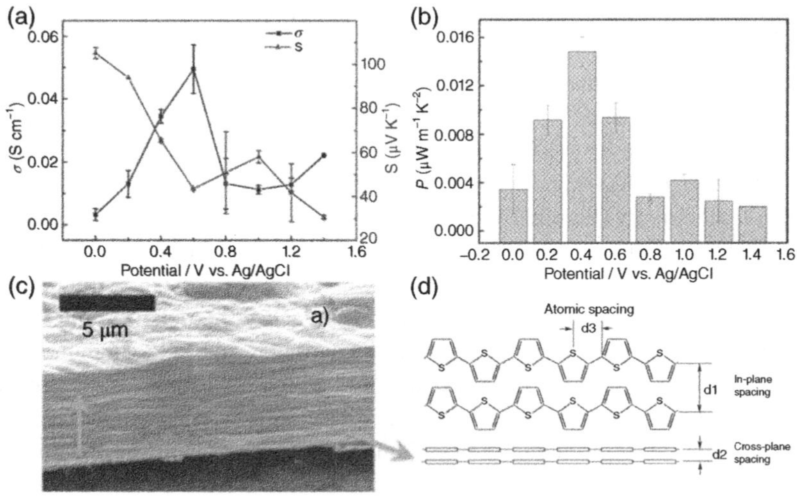

Figure 4.5 (a) Electrical conductivity and the Seebeck coefficient as well as the power factor (b) dependence on the applied potentials for PTFu films. *Source:* Yao et al. [12]. © Royal Society of Chemistry. (c) The image of PTh film with layer structure electrodeposited in $BF_3 \bullet Et_2O$ solution with and without the surfactants of 0.5 mM SDBS, (d) Illustration of PTh molecular packing. *Source:* Santoso et al. [45]. American Chemical Society.

while the Seebeck coefficient increases quasi-linearly ($9\,\mu V\,K^{-1}$ at $V_G = 0\,V$), as shown in Figure 4.6a and b, eventually resulting in the increase of the power factor (up to ~$0.4\,mV\,K^{-1}$ at $V_G = 1.4\,V$) from −0.2 to 0.6 V. Figure 4.6c exhibits the increasing Seebeck coefficient as a function of charge carrier concentration, simultaneously revealing the results of the corresponding increase in the electrical conductivity. Their evolution can also be explained based on the localized electronic levels by the DOS in Figure 4.6d and probability distribution of the charge carriers. Noted that, the mixing states with polaron and bipolaron are always in presence in the disordered PEDOT:PSS (Figure 4.6e). The OECT as a versatile tool has provided clear evidence for the optimization of thermoelectric performance of CPs by the doping control.

4.7 Conclusions

For a thermoelectric plastic film, the electrochemistry has been demonstrated to be a powerful tool that not only allows to fabricate a free-standing polymer film but also precisely controls the actual doping level. Many works

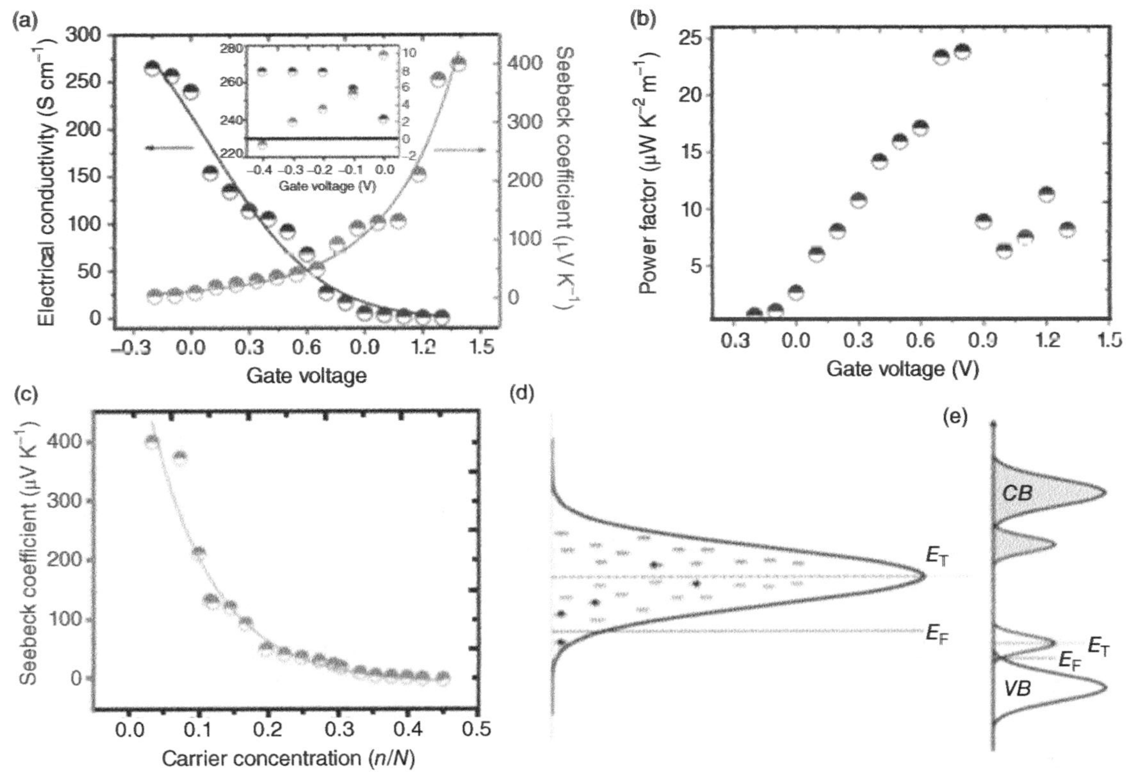

Figure 4.6 (a) Electrical conductivity and the Seebeck coefficient with (b) the estimated power factor dependence on gate voltage. (c) Seebeck coefficient as a function of charge carrier concentration, (d) Gaussian DOS, (e) CB and VB with localized bands at the forbidden energy gap for PEDOT:PSS. *Source:* Bubnova et al. [43]. © American Chemical Society.

have been focused on the optimization of thermoelectric performance for CPs with the remarkably improved results. Although electrochemistry has many advantages, it requires some unique features for CPs with the environmental stability and solubility including the solvent-electrolyte system. Further taking into account the thermoelectric performance, it is desired to attract more attention in the microstructure of polymer chains, the length of conjugated chains, and the type of dopants by electrochemistry.

Acknowledgments

This work was supported by the financial support of the National Natural Science Foundation of China (51762018, 51863009, 52073128, and 22065013), and the Natural Science Foundation of Jiangxi Province (20202ACBL204005, 20202ACBL214005, and 20203AEI003).

References

1 a) Kaloni, T.P., Giesbrecht, P.K., Schreckenbach, G., and Freund, M.S. (2017). *Chem. Mater.* 29: 10248. b) Pathiranage, T.M.S.K., Dissanayake, D.S., Niermann, C.N. et al. (2017). *J. Polym. Sci. Part A: Poly. Chem.* 55: 3327.

2 a) Groenendaal, L., Zotti, G., Aubert, P.-H. et al. (2003). *Adv. Mater.* 15: 855. b) Heinze, J., Frontana-Uribe, B.A., and Ludwigs, S. (2010). *Chem. Rev.* 110: 4724. c) Janáky, C. and Rajeshwar, K. (2015). *Prog. Polym. Sci.* 43: 96.

3 Bufon, C.C.B., Vollmer, J., Heinzel, T. et al. (2005). *J. Phys. Chem. B* 109: 19191.

4 Li, C., Bai, H., and Shi, G. (2009). *Chem. Soc. Rev.* 38: 2397.

5 a) Liu, X., Hu, Y., Shen, L. et al. (2018). *Electrochim. Acta* 288: 52. b) Hu, Y., Hu, D., Ming, S. et al. (2016). *Electrochim. Acta* 189: 64. c) Hu, Y., Wang, Z., Lin, K. et al. (2016). *J. Polym. Sci. Part A: Poly. Chem.* 54: 1583.

6 a) Shi, G., Xu, J., and Fu, M. (2002). *J. Phys. Chem. B* 106: 288.b) Hu, Y., Zhu, D., Zhu, Z. et al. (2016). *ChemPhysChem* 17: 2256.c) Hu, Y., Liu, X., Jiang, F. et al. (2017). *J. Phys. Chem. B* 121: 9281.

7 Roncali, J., Garnier, F., Garreau, R., and Lemaire, M. (1987). *J. Chem. Soc., Chem. Commun.*: 1500.

8 a) Heywang, G. and Jonas, F. (1992). *Adv. Mater.* 192: 116. b) Jonas, F. and Schrader, L. (1991). *Synth. Met.* 41: 831.

9 a) Walczak, R.M. and Reynolds, J.R. (2006). *Adv. Mater.* 18: 1121. b) Walczak, R.M., Jung, J.-H., Cowart, J.S., and Reynolds, J.R. (2007). *Macromolecules* 40: 7777. c) Schottland, P., Zong, K., Gaupp, C.L. et al. (2000). *Macromolecules* 33: 7051.d) Patra, A. and Bendikov, M. (2010). *J. Mater. Chem.* 20: 422.

10 Hu, Y., Jiang, F., Lu, B. et al. (2017). *Electrochim. Acta* 228: 361.

11 Xu, J., Wei, Z., Du, Y., and Pu, S. (2007). *Mater. Lett.* 61: 2486.

12 Yao, W., Shen, L., Liu, P. et al. (2020). *Mater. Chem. Front.* 4: 597.

13 Shi, G., Jin, S., Xue, G., and Li, C. (1995). *Science* 267: 994.

14 Xu, J.K., Shi, G.Q., Qu, L.T., and Zhang, J.X. (2003). *Synth. Met.* 135–136: 221.

15 Yakuphanoglu, F., Liu, H., and Xu, J. (2007). *J. Phys. Chem. B* 111: 7535.

16 Ma, M., Guo, L., Anderson, D.G., and Langer, R. (2013). *Science* 339: 186.

17 Bubnova, O. and Crispin, X. (2012). *Energy Environ. Sci.* 5: 9345.

18 a) Zade, S.S. and Bendikov, M. (2007). *J. Phys. Chem. C* 111: 10662. b) Colella, N.S., Zhang, L., McCarthy-Ward, T. et al. (2015). *Phys. Chem. Chem. Phys.* 17: 26525.

19 Rittmeyer, S.P. and Gross, A. (2012). *Beilstein J. Nanotechnol.* 3: 909–919.

20 The Nobel Prize in Chemistry (2000). Nobel Media AB 2021. http://NobelPrize.org (accessed 26 March 2021).

21 Bubnova, O., Khan, Z.U., Wang, H. et al. (2014). *Nat. Mater.* 13: 190.

22 a) Eckhardt, H., Shacklette, L.W., Jen, K.Y., and Elsenbaumer, R.L. (1989). *J. Chem. Phys.* 91: 1303. b) Li, Y., Cao, Y., Gao, J. et al. (1999). *Synth. Met.* 99: 243.

23 Li, Y., Wen, J., Liu, J. et al. (2011). *Prog. Chem.* 23: 2215.

24 Muñoz, W.A., Singh, S.K., Franco-Gonzalez, J.F. et al. (2016). *Phys. Rev. B* 94: 205202.

25 Milczarek, G. and Inganäs, O. (2012). *Science* 335: 1468.

26 Bharti, M., Singh, A., Samanta, S., and Aswal, D.K. (2018). *Prog. Mater. Sci.* 93: 270.

27 a) Kroon, R., Mengistie, D.A., Kiefer, D. et al. (2016). *Chem. Soc. Rev.* 45: 6147. b) Liu, C., Wang, K., Gong, X., and Heeger, A.J. (2016). *Chem. Soc. Rev.* 45: 4825.

28 a) Jiang, F., Xu, J., Lu, B. et al. (2008). *Chin. Phys. Lett.* 25: 2202. b) Lu, B., Liu, C., Lu, S. et al. (2010). *Chin. Phys. Lett.* 27: 057201.

29 Shi, H., Liu, C., Xu, J. et al. (2013). *ACS Appl. Mater. Interfaces* 5: 12811.

30 Ye, G., Xu, J., Ma, X. et al. (2017). *Electrochim. Acta* 224: 125.

31 a) Patel, S.N., Glaudell, A.M., Kiefer, D., and Chabinyc, M.L. (2016). *ACS Macro Lett.* 5: 268. b) Kang, K., Watanabe, S., Broch, K. et al. (2016). *Nat. Mater.* 15: 896.

32 Yue, R., Chen, S., Lu, B. et al. (2010). *J. Solid State Electrochem.* 15: 539.

33 Hu, Y., Shi, H., Song, H. et al. (2013). *Synth. Met.* 181: 23.

34 Lu, B., Chen, S., Xu, J., and Zhao, G. (2013). *Synth. Met.* 183: 8.

35 Kim, B., Shin, H., Park, T. et al. (2013). *Adv. Mater.* 25: 5483.

36 a) Jia, Y., Li, X., Jiang, F. et al. (2017). *J. Poly. Sci. Polym. Phys.* 55: 1738. b) Jia, Y., Liu, C., Liu, J. et al. (2019). *J. Poly. Sci. Polym. Phys.* 57: 257. c) Jia, Y., Shen, L., Liu, J. et al. (2019). *J. Mater. Chem. C* 7: 3496.

37 a) Wakim, S., Aïch, B.R., Tao, Y., and Leclerc, M. (2008). *Polym. Rev.* 48: 432. b) Aïch, R.B., Blouin, N., Bouchard, A., and Leclerc, M. (2009). *Chem. Mater.* 21: 751.
38 a) Jiang, F., Ren, F., Zhou, W. et al. (2012). *Fuel* 102: 560. b) Xu, J., Zhou, W., Peng, H. et al. (2006). *Polym. J.* 38: 369.
39 Ming, S., Zhen, S., Lin, K. et al. (2015). *J. Electron. Mater.* 44: 1606.
40 Yue, R., Chen, S., Liu, C. et al. (2011). *J. Solid State Electrochem.* 16: 117.
41 Gu, H., Ming, S., Lin, K. et al. (2016). *J. Electron. Mater.* 46: 3124.
42 Yue, R., Lu, B., Xu, J. et al. (2011). *Polym. J.* 43: 531.
43 Bubnova, O., Berggren, M., and Crispin, X. (2012). *J. Am. Chem. Soc.* 134: 16456.
44 Park, T., Park, C., Kim, B. et al. (2013). *Energy Environ. Sci.* 6: 788.
45 Santoso, H.T., Singh, V., Kalaitzidou, K., and Cola, B.A. (2012). *ACS Appl. Mater. Interfaces* 4: 1697.

5

Thermoelectric Properties of Conducting Polymers with Ionic Conductors

Zeng Fan[1], Jianyong Ouyang[2], and Lujun Pan[1]

[1] School of Physics, Dalian University of Technology, Dalian, P. R. China
[2] Department of Materials Science and Engineering, National University of Singapore, Singapore, Singapore

5.1 Introduction

Current progress achieved on organic thermoelectric (TE) materials is mainly based on conducting polymers which have a conjugated backbone structure in doped states. As representative *p*-type organic semiconductors, poly(3,4-ethylenedioxythiophene) (PEDOT), polyaniline (PANi), polypyrrole (PPy), and their derivatives have been receiving much attention. Based on the type of charge carriers, these conducting polymers are basically electronic conductors or exhibit electronic TE behaviors, which have high electrical conductivities with relatively low Seebeck coefficients (commonly less than $1\,\mathrm{mV\,K^{-1}}$). According to the definition of figure-of merit $ZT = S^2\sigma T/\kappa$ (where S, σ, T and κ represent the Seebeck coefficient, electrical conductivity, absolute temperature, and thermal conductivity, respectively), high-efficient TE materials should meet the criteria of high Seebeck coefficient, high electrical conductivity, and low thermal conductivity at the same time. However, due to the limitations arising from the band structure and effect of electron–phonon coupling, strong trade-offs always exist between the Seebeck coefficient-electrical conductivity and electrical conductivity–thermal conductivity in these electronic conductors, eventually making their *ZT* optimization a rather challenging task. Recently, electrolyte-based thermal energy conversion has become noticeable due to their ultrahigh Seebeck coefficients which are commonly in the range of $\mathrm{mV\,K^{-1}}$ [1]. But their low ionic conductivities ($<0.1\,\mathrm{S\,cm^{-1}}$) unfortunately hinder the

Flexible Thermoelectric Polymers and Systems, First Edition. Edited by Jianyong Ouyang.

development of ionic heat-to-electricity conversion toward high-performance practical applications.

In principal, mixed ionic-electronic conductors (MIECs) that conduct both ions and electrons should also hold great promise for TE applications. In previous studies, the MIECs have been intensively investigated as catalysts, separation membranes, and electrodes for electrochemical devices. However, their applications in terms of energy conversion have been neglected for a long time until their unique TE characteristics are discovered. Over the past few years, the integration of ionic conductors with conducting polymers is emerging as a new research direction for developing low-cost, flexible, and all-solution-processable TE organics; a room-temperature *ZT* comparable with that of the conventional inorganic bismuth telluride (Bi_2Te_3) has been even achieved on the heterostructures constructed by ionic conductors and conducting polymers. These exciting advances may suggest a new avenue and a bright future for the development of high-performance TE polymers based on the conducting polymers containing ionic charge carriers.

This chapter provides an introduction on the fundamentals and TE characteristics of the ionic–electronic coupled organics and some novel outstanding ion-conducting polymers. At the last part, the application prospects of these novel ion-containing organic TE materials are also briefly discussed.

5.2 Mixed Ionic-Electronic Conductors

Organic MIECs mainly refer to polymers that are intrinsically soft electrical (semi-)conductors but allow easy solvation or transport of ionic species. Figure 5.1 illustrates the chemical structures of many representative organic MIECs [2]. Typically, these polymer-based organic MIECs are based on intrinsically conducting polymers which have π-conjugated backbones, and in the meanwhile contain (Figure 5.1 [**I, III** and **V**]) or allow easy solvation of mobile ionic species (Figure 5.1 [**II, IV,** and **VI**]). Alternatively, depending on whether there are segregations of predominantly ion-conducting and electron-conducting regions, the organic MIECs can be categorized into either heterogeneous (Figure 5.1a and b) or homogenous (Figure 5.1c) systems. Owing to their low thermal conductivity and complex conducting modes, the organic MIECs have been explored as novel TE materials. Of particular interest is the poly(3,4-ethyelenedioxythiophene):poly(styrenesu lfonate) (PEDOT:PSS), which is a heterogeneous blend of electrically conducting conjugated PEDOT and ionically conducting polyelectrolyte poly(4-styrenesulfonic acid) (PSSH).

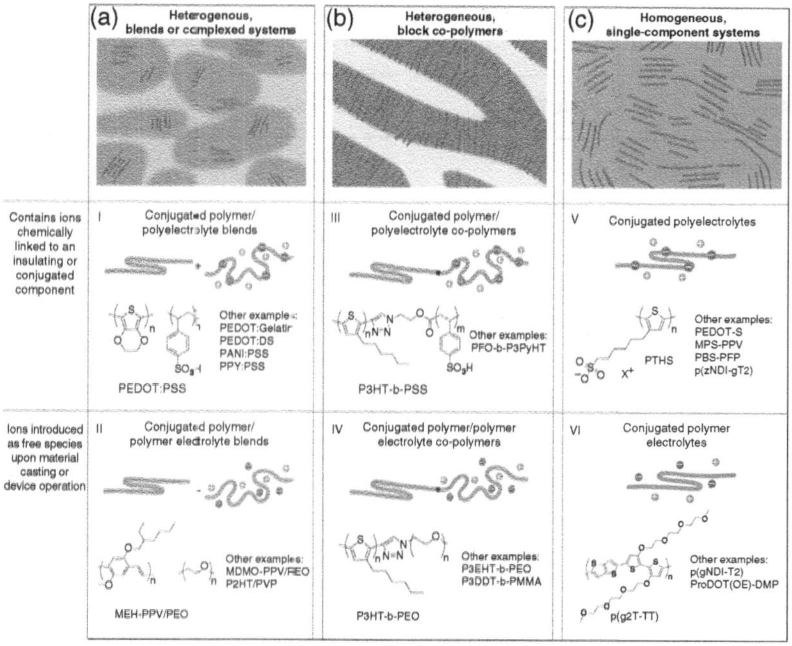

Figure 5.1 Classes of organic MIECs. (a–c) MIECs categorized between heterogeneous and homogeneous systems. (a) Heterogeneous, blends or complex systems, (b) heterogeneous block copolymers and (c) homogeneous single-component systems. (I, III, and V) MIECs intrinsically containing ionic charges. (I) Conjugated polymer/polyelectrolyte blends, (III) conjugated polymer/polyelectrolyte copolymers, and (V) Conjugated polyelectrolytes. (II, IV, and VI) MIECs with free ionic charges induced upon material preparation or device operation. (II) Conjugated polymer/polymer electrolyte blends, (IV) conjugated polymer/polymer electrolyte copolymers and (VI) conjugated polymer electrolytes. *Source:* Paulsen et al. [2]. © Springer Nature.

Despite that the ZT of the MIECs is commonly low (<0.05) at the ambient conditions, Crispin's team first discovered that the MIECs can exhibit significantly different TE behaviors from those of the electronic and ionic conductors [1b, c]. A comparative study in terms of the TE characteristics was carried out between a pure electronic conductor (PEDOT:toyslate = PEDOT:Tos), a pure ionic conductor (poly(sodium 4-styrenesulfonate) = PSSNa), and a MIEC (PEDOT:PSS), as listed in Table 5.1. In these three polymeric conductors, the electrical charge carriers are h^+ (holes), Na^+ (cations), and both h^+ and H^+ (holes and cations), respectively. Since the ions are readily dissociated at an elevated relative humidity (RH) level but the holes not, the TE properties of the pure electronic conductor, pure ionic conductor, and MIEC in

Table 5.1 Chemical structures and charge carriers in typical a pure electronic conductor (PEDOT:Tos), a pure ionic conductor (PSSNa), and a MIEC (PEDOT:PSS).

Polymer	PEDOT:Tos	PEDOT:PSS	PSSNa
Chemical structure	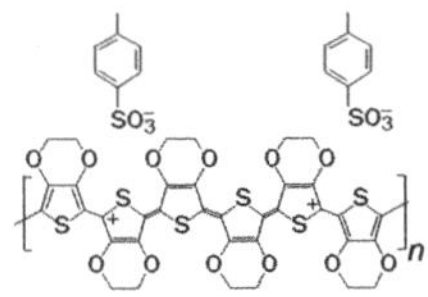	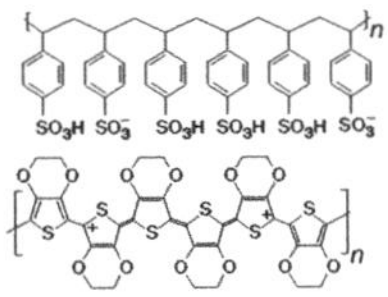	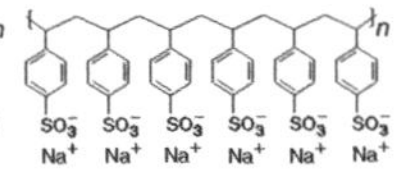
Charge carriers	Holes (h^+)	Holes (h^+) Cations (H^+)	Cations (Na^+)

Source: Adapted with permission Ail et al. [1c]. © John Wiley & Sons.

consequence show distinguishable dependences on the RH level. As shown in Figure 5.2a, the electrical conductivity of the electronic conductor is insensitive to RH, while that of the pure ionic conductor exhibits an exponential increase as a function of RH. When the RH level elevates from 10 to 80%, the electrical conductivity of the MIEC first decreases and then increases. This trend could be attributed to the competition between the swelling effect of the hygroscopic PSS chains and the enhanced mobility of H^+ ions by water dissociation.

At a temperature gradient $\Delta T = 1\,K$, the ratio of open-circuit voltage $(V_{OC})/\Delta T$ defines the Seebeck coefficient. For these three categories of polymeric conductors, the variations of their V_{OC} with time also possess distinct characteristics, as displayed in Figure 5.2b. Upon ΔT, the V_{OC} of the electronic conductor increases rapidly within ≈30 seconds and then keeps constant at several tens of $\mu V\,K^{-1}$, while that of the pure ionic counterpart can be in the $mV\,K^{-1}$ magnitude but the evolution time is up to ~100 minutes. For the MIEC, the variation of its V_{OC} with time delivers a more complex feature. The V_{OC} first instantly increases with ΔT, but starts to decrease before the ΔT stabilization time. The V_{OC} peak value that the MIEC achieves, shows a strong dependence on the RH level, as shown in Figure 5.2c. The presence of this V_{OC} peak may indicate that the charge transport in the MIEC could arise from both the electronic and ionic contributions. Based on an analysis of its V_{OC} evolution curve, the mechanism for charge transport in an MIEC is hypothetically proposed, as shown in Figure 5.2d. In stage (i), the holes showing instant response to ΔT first dominate the charge transport and establish an internal electric field due to the Seebeck effect. After the distribution of holes tends to stabilize, the immigration of large ions becomes pronounced (stage (ii)). The thermodiffusion of ions upon ΔT

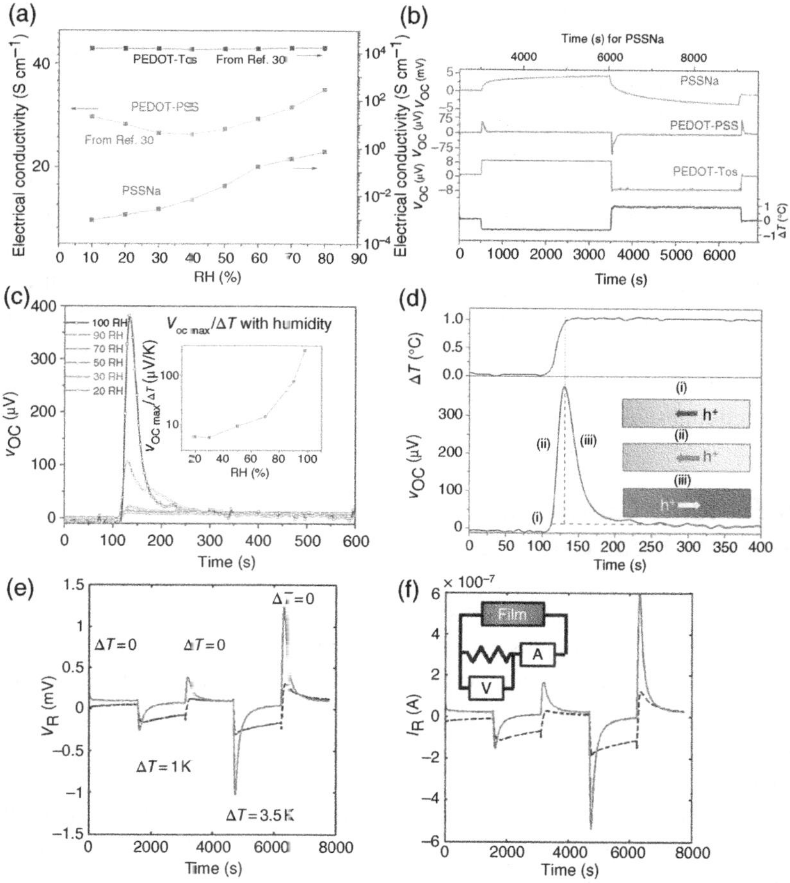

Figure 5.2 TE properties of MIECs. Comparisons of (a) electrical conductivity and (b) V_{OC}-time dependence of three polymeric conductors, i.e. a pure electronic conductor, a pure ionic conductor and a MIEC. (c) V_{OC}-time dependence of PEDOT:PSS as a function of RH. (d) Schematic diagram of the mechanism for explaining the V_{OC}-time dependence of the MIEC. *Source:* Ail et al. [1c]. © John Wiley & Sons. (e) V_{OC}-time and (f) current-time dependences of PEDOT:PSS (solid line) and PEDOT:PSS (dash line) with silver addition. *Source:* Chang et al. [3]. © American Chemistry Society.

is called the Soret effect. Compared with the V_{OC} generated by electronic conductors due to the Seebeck effect, the V_{OC} arising from the Soret effect by ionic contribution can be larger by several orders of magnitude, therefore giving rise to the high V_{OC} value. In stage (iii), as the ion concentration further increases and becomes sufficiently large, an internal electric filed forms and leads to a drift current of the holes toward the opposite direction. This process weakens the Soret effect by ions, thus resulting in a significantly reduced V_{OC} value. The duration of the Seebeck coefficient enhancement induced by ions is related to whether there is an electrochemical reaction occurring at the MIEC/electrode interface [3]. For the interface without the occurrence of electrochemical reaction (e.g. PEDOT:PSS and gold electrodes), the decline of V_{OC} is in an exponential manner, whereas for that with an electrochemistry (e.g. PEDOT:PSS added with silver and silver electrodes), the time for the V_{OC} decay can be much longer (Figure 5.2e and f). The prolonged V_{OC} decay time is attributed to the generation and consumption of active electrode ions during the thermodiffusion process. Since the TE properties of the MIECs possess a high sensitivity to RH, they can be a promising candidate as the sensing medium for temperature-humidity multifunctional sensing applications.

5.3 Ionic Conductor/Conducting Polymer Heterostructures

Although the pristine PEDOT:PSS is a typical MIEC, the PEDOT:PSS films after various posttreatments commonly exhibit an electronic TE behavior due to the effective depletion of excessive PSS chains [1b, 4]. The TE properties of conducting polymers, particularly the PEDOT:PSS films posttreated with secondary dopants, acids and reducing agents have been widely studied [5]. Power factors exceeding $300\,\mu W\,m^{-1}\,K^{-2}$ have been reported for the PEDOT:Tos films treated with reducing agent tetrakis(dimethylamino) ethylene (TDAE) and the PEDOT:PSS films sequentially treated with sulfuric acid and sodium hydroxide (NaOH) [6]. Interestingly, Ouyang's team found that the TE properties of the conducting polymer could be further enhanced remarkably by ion accumulation of an ionic conductor on the polymer surface [7]. Novel ionic conductor/conducting polymer heterostructures are prepared by simply spin-coating the ionic liquid (IL)-methanol solution on the surface of the acid–base-treated PEDOT:PSS (AB-PEDOT:PSS) film. As shown in Figure 5.3a and b, compared with the conducting polymer film counterpart, the heterostructure constructed by the ionic liquid (IL) 1-ethyl-3-methylimidazolium dicyanamide (EMIM-DCA) and AB-PEDOT:PSS successfully achieves a twofold enhancement in its Seebeck coefficient while the electrical conductivity only slightly decreases. As a result, the corresponding power factor reaches up to $754\,\mu W\,m^{-1}\,K^{-2}$ (Figure 5.3c). Considering that the thermal conductivities of the ionic liquid and conducting polymer are both in the range of $0.2–0.5\,W\,m^{-1}\,K^{-1}$, the calculated ZT of the heterostructure is up to 0.75, which is even comparable with that of a conventional inorganic TE material bismuth telluride (Bi_2Te_3) at room

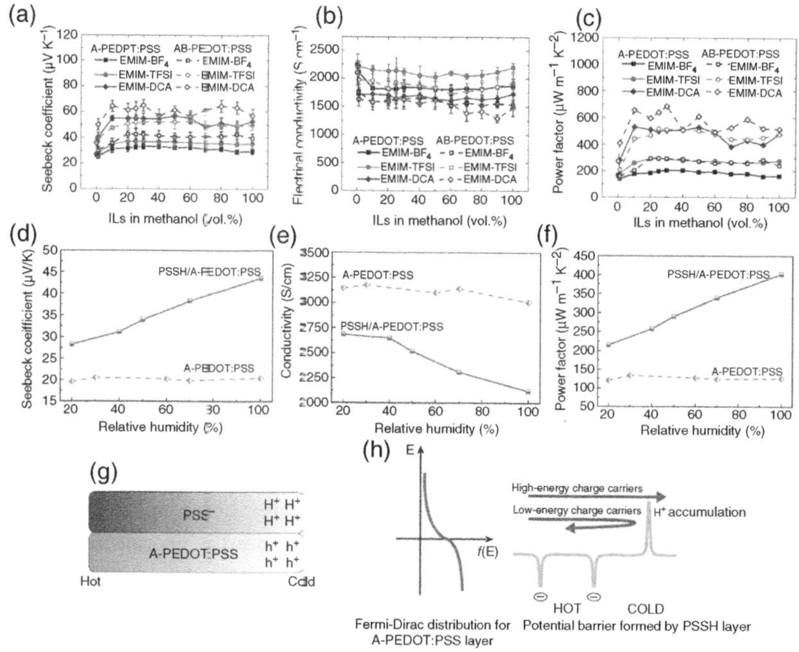

Figure 5.3 TE properties of ionic conductor/conducting polymer heterostructures and the corresponding mechanism for Seebeck coefficient enhancement. (a) Seebeck coefficient, (b) electrical conductivity, and (c) power factor of IL/PEDOT:PSS heterostructures. *Source:* Fan et al. [7]. © Elsevier. (d) Seebeck coefficient, (e) electrical conductivity and (f) power factor of PSSH/PEDOT:PSS heterostructures. (g and h) Schematic diagram of the mechanism for explaining the Seebeck coefficient enhancement in an ionic conductor/conducting polymer heterostructure. (Left) Fermi-dirac distribution of charge carriers. (Right) Hole potential at the PEDOT:PSS surface. A-PEDOT:PSS represents the PEDOT:PSS film treated with sulfuric acid. *Source:* Guan et al. [8]. © Royal Society of Chemistry.

temperature. In addition to EMIM-DCA, the remarkable Seebeck coefficient enhancement is also observed on the heterostructures constructed by AB-PEDOT:PSS and another two ILs 1-ethyl-3-methylimidazolium tetrafluoroborate($EMIM\text{-}BF_4$) and 1-ethyl-3-methylimidazolium bis(trifluoromethylsulfonyl) imide (EMIM:TFSI). At a concentration higher than 30 vol.% in methanol, the IL fully covers the surface of the PEDOT:PSS film. The TE properties of these IL/PEDOT:PSS heterostructures become insensitive to the IL concertation. Therefore, the greatly enhanced Seebeck coefficient of the conducting polymer by ILs presumably arises from an interfacial effect induced by ion accumulation.

A similar effect has also been observed on the PEDOT:PSS films coated with a layer of solid polyelectrolytes, e.g. PSSH and PSSNa [8]. Since the dissociations of protons and Na^+ ions in the solid PSSH and PSSNa are promoted in the presence of water, the TE properties of the PSSH/PEDOT:PSS and PSSNa/PEDOT:PSS heterostructures exhibit a strong relevance with the RH levels (Figures 5.3d–f). Without the PSSH layer, the neat PEDOT:PSS film has a Seebeck coefficient of 18.9 $\mu V K^{-1}$ and an electrical conductivity of 3152 $S cm^{-1}$, both of which are insensitive to RH. After being coated with a layer of PSSH, the PEDOT:PSS film achieves significant enhancements in the Seebeck coefficient and only slightly decreases in the electrical conductivity. The effect becomes more pronounced at an elevating RH level. But owing to the remarkably enhanced Seebeck coefficient over the slightly reduced electrical conductivity, the power factor of the PSSH/PEDOT:PSS heterostructure increases from 215.2 $\mu W m^{-1} K^{-2}$ at 20% RH to 401 $\mu W m^{-1} K^{-2}$ at 100% RH, which is much higher than that reported for the PEDOT-based materials prepared by other approaches [5b, 9].

For these ionic conductor/conducting polymer heterostructures, the mechanism for the Seebeck coefficient enhancement is proposed. For an ionic conductor/conducting polymer heterostructure, charge thermodiffusion takes place on both the holes and ions upon ΔT (Figure 5.3g). As is known, the ion immigration arising from the Soret effect can produce a much larger thermovoltage than the hole transport arising from the Seebeck effect. The accumulated ions in the ionic conductor layer, therefore, creates a potential barrier for the hole transport in the conducting polymer film. In this case, only the holes with high energy contribute to the charge transport, while those with lower energy are filtered out. The equivalent effect is to increase the mean hole energy at the cold side. According to the expression of Seebeck coefficient: $S = (E_F - E_T)/T$ (where E_F, E_T and T are the Fermi energy, mean charge energy, and absolute temperature, respectively), an increase in the mean energy of charge carriers directly enhances the Seebeck effect in the conducting polymer (Figure 5.3h).

For the aforementioned ionic conductor/conducting polymer heterostructures based on PEDOT:PSS, the presence of interaction between the ionic conductor (i.e. IL and PSSH) and PEDOT:PSS is evidenced by the PEDOT conformation change based on the in situ Raman spectra under ΔT (Figure 5.4). The Raman band located at 1400–1500 cm^{-1} corresponds to the $C_\alpha = C_\beta$ stretching [10]. For both the PEDOT:PSS film and PSSH/PEDOT:PSS heterostructure, this Raman band exhibits a ΔT-induced red-shift. However, the shift in the latter is more salient than that in the former, which indicates the PEDOT conformational transition from benzoid to quinoid on the condition of ion accumulation in the ionic

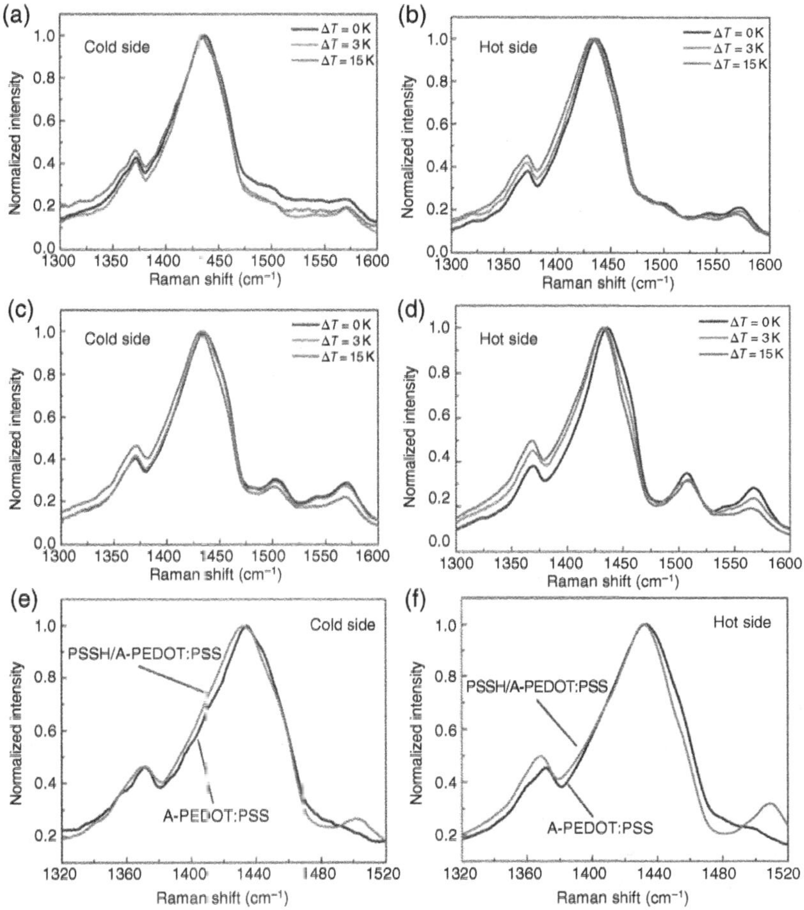

Figure 5.4 In-situ Raman spectra of the ionic conductor/conducting polymer heterostructure. (a) Cold side and (b) hot side of the neat PEDOT:PSS film. (c) Cold side and (d) hot side of the PSSH/PEDOT:PSS heterostructure. Comparisons of the Raman spectra of the PSSH/PEDOT:PSS heterostructure and neat PEDOT:PSS film at (e) cold side and (f) hot side at $\Delta T = 15$ K. *Source:* Guan et al. [8]. © Royal Society of Chemistry.

conductor (PSSH) layer. Notably, since the improved TE properties in the ionic conductor/conducting polymer heterostructures are results of enhanced Seebeck effect in the conducting polymer, the heterostructure behaves exactly like a pure electronic conductor, for which the V_{OC} not only shows an instant response to ΔT but also remains unchanged after stabilization.

5.4 High-Performance Ion-Conducting TE Polymers

Except for the MIECs and ionic conductor/conducting polymer heterostructures, recently high TE properties in particular ultrahigh Seebeck coefficients exceeding $10\,\text{mV}\,\text{K}^{-1}$ have also been reported for pure ionic conductors [1d, 11]. Upon ΔT, a large ionic Seebeck coefficient originates from a high degree of ion separation and accumulation at the two ends of the sample due to the Soret effect. Table 5.2 lists the TE properties of several representative ion-conducting polymers reported in recent years. It can be observed that the Seebeck coefficients of most pure ionic conductors can be several orders greater than those of the electronic conductors. But considering the low ionic conductivities, their *ZT* values appear to be far from satisfactory for practical applications.

Generally, the Seebeck coefficients of ionic conductors are related to many factors including ion species, ion sizes, concentrations, and viscosities [16]. It is worthwhile pointing out that the ionic Seebeck coefficient is strongly dependent on the interaction between the ions and their transport media in the ionic conductors. Even for an ionic conductor system with fixed compositions, its ionic Seebeck coefficient can be further tailored through a modulation on its ion-solvent interaction. Taking a typical ionic conductor poly(ethylene oxide) (PEO)-NaOH aqueous electrolyte as an example, the Seebeck coefficient of its electrolyte in bulk is ~$10\,\text{mV}\,\text{K}^{-1}$. But after infiltrating it into an ionic selective cellulosic membrane extracted from natural wood, the electrolyte can deliver a Seebeck coefficient of $24\,\text{mV}\,\text{K}^{-1}$ [11a] (Figure 5.5a). The ionic selective cellulosic membrane is prepared through a simple delignification and oxidation treatment to the natural wood. The mechanism for explaining its effect on enhancing the Seebeck coefficient of PEO-NaOH is illustrated in Figure 5.5b. In natural wood, cellulose nanofibers show highly packing alignment along the tree growth direction. The delignification process first produces a specific type of cellulose that can form Na-cellulose complex after the electrolyte infiltration. The oxidation of the resulting cellulose, then, facilitates a high density of cellulose nanofibers with negative charges. In this case, the ion transport is confined at a sub-nanometer scale. Owing to the formation of Na-cellulose complex and the charged and aligned cellulose nanochannels for ion transport, the thermodiffusion of the ions becomes highly selective, thus giving rise to ultrahigh Seebeck coefficient.

A giant Seebeck coefficient up to $26.1\,\text{mV}\,\text{K}^{-1}$ and *ZT* as high as 0.75 have been observed on the ionogels made of ILs and co-polymer poly(vinylidene fluoride-co-hexafluoropropylene) (PVDF-HFP) by Ouyang's team [11b]. Three common ILs, i.e. EMIM:DCA, EMIM:TFSI, and 1-butyl-3-methylimidazolium tetrafluoroborate (BMIM:BF_4) were dissolved in acetone and mixed with

Table 5.2 TE properties of several representative ion-conducting polymers.

Materials	Seebeck coefficient (mVK^{-1})	Electrical conductivity ($S cm^{-1}$)	Thermal conductivity ($W m^{-1} K^{-1}$)	ZT	Reference
TBAN-dodecanol	7	1.2×10^{-5}	0.17 0.001	0.001	[12]
PEO-NaOH	11	8.2×10^{-5}	0.2	0.014	[1d]
PSSNa	0.12	1.2×10^{-2}	0.5	0.013	[1c]
PSSH	8	8×10^{-2}	0.38	0.4	[13]
Nafion	−2	1.2×10^{-3}	—	—	[14]
Graphene oxide	9	4.0×10^{-3}	3	0.004	[15]
Cellulose/PEO-NaOH	24	2.0×10^{-2}	0.48	0.7	[11a]
Ionogels	26	6.3×10^{-3}	0.175	0.75	[11b]

Source: Cheng et al. [11b]. © John Wiley & Sons.

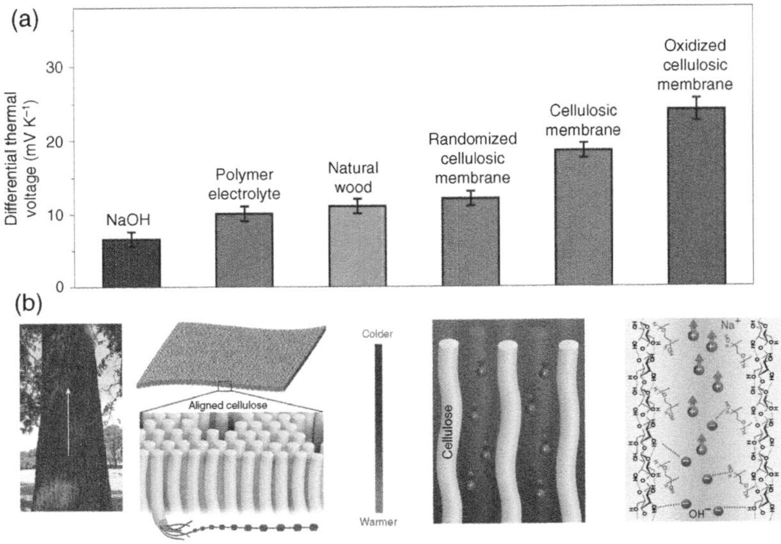

Figure 5.5 Ultrahigh Seebeck coefficient of cellulose ionic conductors and the corresponding mechanism for Seebeck coefficient enhancement. (a) Seebeck coefficient. (b) Schematic diagram of the mechanism for explaining the Seebeck coefficient enhancement in a cellulose ionic conductor. *Source:* Li et al. [11a]. © Springer Nature.

PVDF-HFP (all chemical structures are as shown in Figure 5.6a). The appearance and TE properties of the ionogel films obtained by drop-casting their corresponding mixtures on glass substrates were displayed in Figure 5.6b and c. These ionogel films were flexible, semi-transparent, and in quasi-solid state. All of them exhibited positive Seebeck coefficients in magnitude of mVK^{-1} at room temperature, but the Seebeck coefficient value was reversely correlated with the viscosity of the IL. Apart from the species of Ils, the TE properties of the ionogels strongly depended on the IL loading. As shown in Figure 5.6d–f, the highest power factor of $455\,\mu W\,m^{-1}\,K^{-2}$ with corresponding ZT of 0.75 were achieved, when the EMIM:DCA loading was 80 wt.%. According to the characterizations by Fourier-transform infrared spectroscopy on the neat copolymer and the ionogels (Figure 5.6g and h), electroactive β-phase PVDF-HFP was absent in the former but emerged in the latter, indicating the existence of strong ion–dipole interaction between the PVDF-HFP and ILs. The dipole moment of PVDF-HFP favors the dissociation of cations and anions in ILs. The IL with a low viscosity possesses ions with high freedom, which gives rise to strong interaction with the copolymer and consequently results in a high Seebeck coefficient. Due to the factor that the TE effect of the ionic conductors is predominated by only ion immigration under ΔT, the ion-conducting TE polymers behave as a capacitor.

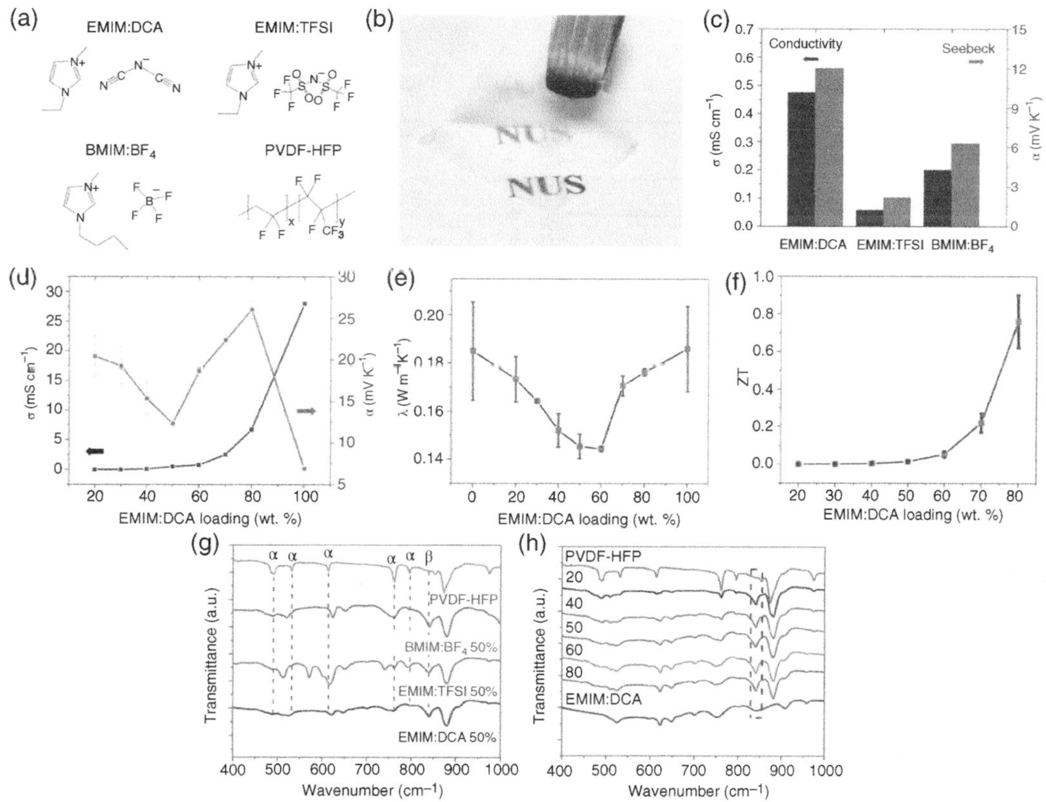

Figure 5.6 TE properties and FT-IR spectra of ionogels. (a) Chemical structures of ILs and copolymer PVDF-HFP. (b) Appearance of an ionogel. (c) Electrical conductivity and Seebeck coefficient of EMIM-DCA/PVDF-HFP, EMIM-TFSI/PVDF-HFP and BMIM-BF_4/PVDF-HFP ionogels. (d) Electrical conductivity and Seebeck coefficient, (e) thermal conductivity, and (f) *ZT* of the EMIM-DCA/PVDF-HFP ionogel as a function of EMIM-DCA loading. FT-IR spectra of (g) PVDF-HFP and three ionogels and (h) EMIM-DCA/PVDF-HFP ionogel as a function of EMIM-DCA loading. *Source:* Cheng et al. [11b]. © John Wiley & Sons.

5.5 Applications of Electronic–Ionic Coupled TE Organics

For widely reported electronic TE materials, TE generators (TEGs) address the most common type of application. With additional TE effects arising from the ionic contribution, the electronic-ionic coupled TE organics suggest wider application prospect.

5.5.1 TE Generators

First, the ionic conductor/conducting polymer heterostructures which show identical TE features as an electronic conductor have been demonstrated as high-performance organic TEGs by Ouyang's team [7]. As shown in Figure 5.7a and b, when the IL/PEDOT:PSS heterostructure is connected with an external resistor and subjected to a ΔT, constant signal can be observed in the output voltage, which further confirms the intrinsically

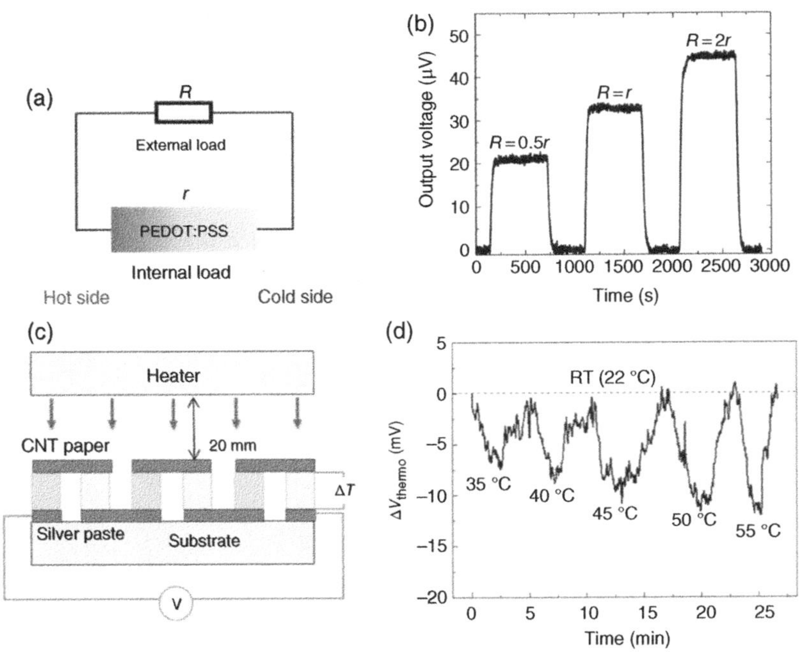

Figure 5.7 Applications of the IL/PEDOT:PSS heterostructure and EMIM-TFSI/PVDF-HFP gel as TEGs. (a) Schematic diagram and (b) output voltage of the IL/PEDOT:PSS heterostructure. *Source:* Fan et al. [7]. © Elsevier. (c) Schematic diagram and (d) output voltage of the EMIM-TFSI/PVDF-HFP gel. *Source:* Zhao et al. [17]. © Springer Nature.

electronic TE behavior of the ionic conductor/conducting polymer heterostructures. Recently, a full ionic TE module consisting of 18 pairs of *p-n* polymer gel TE legs is demonstrated by Crispin's team [17] (Figure 5.7c). The ionic polymer gel is EMIM-TFSI/PVDF-HFP, for which the Seebeck coefficient is found to be variable between −4 and $14\,\mathrm{mV\,K^{-1}}$ with a modulation on the composition. Owing to the gel nature, the TEG is fabricated via screen printing. As shown in Figure 5.7b, the output voltage reaches from 7 to 12 mV when the temperature of the hot side increases from 35 to 55 °C (Figure 5.7d).

5.5.2 Ionic TE Capacitors

Given that the ion-conducting polymers are intrinsically polymer electrolytes, a straight forward application of their heat-to-electricity conversion is to build ionic TE capacitors [1d, 11b, 18]. Figure 5.8a presents the thermally charging–discharging process of an ionic TE capacitor prepared using the ionogel under ΔT. In stage I, a thermovoltage is firstly generated across the two ends of the ionogel due to the Soret effect. By connecting the two electrodes, an electrical current then flows through the external circuit to charge the device (stage II). After charging, the ΔT is removed (stage III). As a consequence, the thermovoltage decays to 0 across the ionogel, but electrons are stored at the electrodes. In this case, when an external load is connected again, the device discharges as a conventional capacitor with

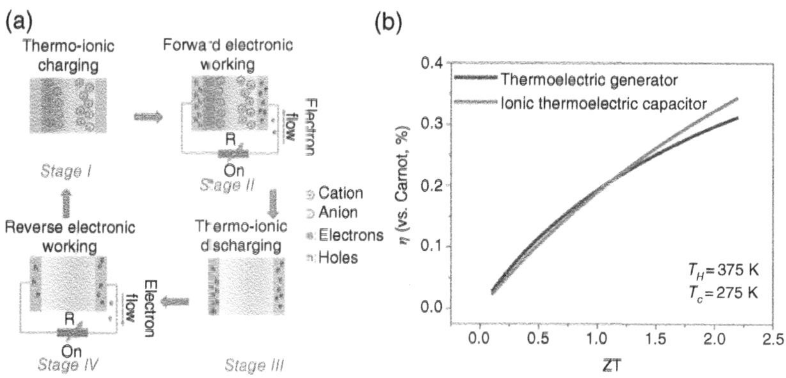

Figure 5.8 Working principal of an ionic capacitor and a comparison of conversion efficiency between an ionic capacitor and a TEG. (a) Working principal of an ionic capacitor and (b) the calculated conversion efficiencies of an ionic capacitor and a TEG as a function of *ZT*. The temperature at hot and cold sides were assumed to be 375 and 275 K, respectively. *Source:* Cheng et al. [11b]. © John Wiley & Sons.

electrons flowing away from the electrodes (stage IV). A calculation on the conversion efficiency predicts that, at a *ZT* exceeding 1.5, an ionic TE capacitor could be even more efficient than a TEG in terms of the heat-to-electricity conversion (Figure 5.8b).

5.5.3 Multifunctional Sensors

Based on a novel MIEC aerogel, a multifunctional sensor that can simultaneously or consecutively detect an integrated pressure–temperature-RH stimulus with decoupled signals is developed [19]. The MIEC aerogel is prepared by freeze-drying the mixture of a MIEC PEDOT:PSS, aerogel forming skeleton nanofibrillated cellulose (NFC) and a cross-linking agent glycidoxypropyl trimethyoxysilane (GOPS) (Figure 5.9a). The MIEC aerogel is shaped into a branch-like configuration to ensure a fast response to the swift variation in humidity (Figure 5.9b). Owing to the highly porous structure of the MIEC aerogel, the pressure applied on the sensor is reflected by a change in its resistance (i.e. the slope of the current–voltage curve in Figure 5.9c). The unique Seebeck coefficient feature of the MIEC is the key factor that facilitates the

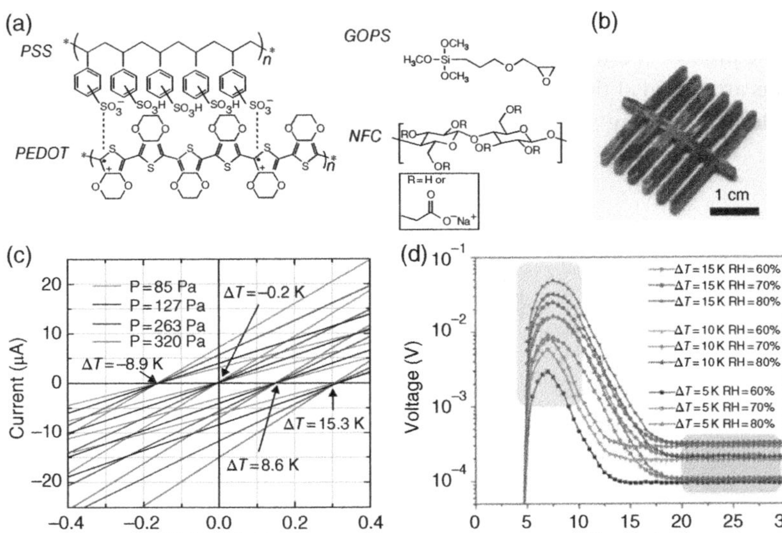

Figure 5.9 Application of a MIEC aerogel as pressure–temperature-humidity multifunctional sensor. (a) Chemical structures of PEDOT:PSS, GOPS and NFC. (b) Appearance of the MIEC aerogel-based multifunctional sensor in a branch-like shape. (c) *I*–*V* curves at different pressures and ΔT and (d) V_{OC}-time dependences of the MIEC aerogel-based multifunctional sensor. *Source:* Han et al. [19]. © John Wiley and Sons.

independent measurement of temperature and RH. As shown in Figure 5.9d, the Seebeck effect at steady state is in accordance with the ΔT due to the Seebeck effect, while the peak value can provide information on the RH level by the Soret effect. As a result, by measuring the current–voltage curve and V_{OC}-time dependence, the sensor enables decoupling of three individual signals, i.e. pressure, temperature, and RH signals from a complex input.

5.6 Summary

Ionic conductors with ultrahigh ionic Seebeck coefficients in the magnitude of $mV K^{-1}$ have attracted great attention in recent years. Although their electrical conductivities are much lower than those of the common electronic conductors, the incorporation of the ionic conductors with electron-conducting polymers interestingly brings out new observations on the TE characteristics. Until now, ultrahigh ZT values of 0.75 have been achieved on the ionic liquid/PEDOT:PSS heterostructures and IL/copolymer ionogels. Moreover, the novel electronic–ionic coupled TE organics broaden the application prospects of all-organic-based TE materials. Novel applications such as ionic TE capacitors and multifunctional sensors have successfully been demonstrated on the ionic conductors and MIECs, respectively. All these exciting achievements made on the ion-containing TE materials may open up a new research direction toward the development of high-performance TE organics.

Acknowledgments

This work was supported by the National Natural Science Foundation of China (grant numbers 51803018, 51972039, and 51661145025); Liaoning Revitalization Talents Program (grant number XLYC1902122), and the Fundamental Research Funds for the Central Universities (grant number DUT18RC(3)010).

References

1 a) Malti, A., Edberg, J., Granberg, H. et al. (2015). *Adv. Sci.* 3: 1500305. b) Wang, H., Ail, U., Gabrielsson, R. et al. (2015). *Adv. Energy Mater.* 5: 1500044. c) Ail, U., Jafari, M.J., Wang, H. et al. (2016). *Adv. Funct. Mater.* 26: 6288. d) Zhao, D., Wang, H., Khan, Z.U. et al. (2016). *Energ. Environ. Sci.* 9: 1450. e) Jiao, F., Naderi, A., Zhao, D. et al. (2017). *J. Mater. Chem. A* 5: 16883.

2 Paulsen, B.D., Tybrandt, K., Stavrinidou, E., and Rivnay, J. (2020). *Nat. Mater.* 19: 13.
3 Chang, W.B., Fang, H., Liu, J. et al. (2016). *ACS Macro Lett.* 5: 455.
4 Bubnova, O., Khan, Z.U., Wang, H. et al. (2014). *Nat. Mater.* 13: 190.
5 a) Yao, H., Fan, Z., Cheng, H. et al. (2018). *Macromol. Rapid Commun.*: 1700727. b) Zhang, S., Fan, Z., Wang, X. et al. (2018). *J. Mater. Chem. A* 6: 7080. c) Lee, S.H., Park, H., Kim, S. et al. (2014). *J. Mater. Chem. A* 2: 7288. d) Chen, Y., Zhao, Y., and Liang, Z. (2015). *Energ. Environ. Sci.* 8: 401.
6 a) Fan, Z., Li, P., Du, D., and Ouyang, J. (2017). *Adv. Energy Mater.* 7: 1602116. b) Bubnova, O., Khan, Z.U., Malti, A. et al. (2011). *Nat. Mater.* 10: 429.
7 Fan, Z., Du, D., Guan, X., and Ouyang, J. (2018). *Nano Energy* 51: 481.
8 Guan, X., Cheng, H., and Ouyang, J. (2018). *J. Mater. Chem. A* 6: 19347.
9 a) Lee, H.J., Anoop, G., Lee, H.J. et al. (2016). *Energy Environ. Sci.* 9: 2806. b) Zhang, L., Deng, H., Liu, S. et al. (2015). *RSC Adv.* 5: 105592. c) Zaia, E.W., Sahu, A., Zhou, P. et al. (2016). *Nano Lett.* 16: 3352. d) Wang, C., Sun, K., Fu, J. et al. (2018). *Adv. Sustain. Syst.* 2: 1800085.
10 Ouyang, J., Xu, Q., Chu, C.-W. et al. (2004). *Polymer* 45: 8443.
11 a) Li, T., Zhang, X., Lacey, S.D. et al. (2019). *Nat. Mater.* 18: 608. b) Cheng, H., He, X., Fan, Z., and Ouyang, J. (2019). *Adv. Energy Mater.* 9: 1901085.
12 Bonetti, M., Nakamae, S., Roger, M., and Guenoun, P. (2011). 114513. *The Journal of chemical physics*: 134.
13 Kim, S.L., Lin, H.T., and Yu, C. (2016). *Adv. Energy Mater.* 6: 1600546.
14 Chang, W.B., Evans, C.M., Popere, B.C. et al. (2016). *ACS Macro Lett.* 5: 94.
15 Kim, S.L., Hsu, J.-H., and Yu, C. (2018). *Nano Energy* 48: 582.
16 a) Laux, E., Uhl, S., Journot, T. et al. (2016). *J. Electron. Mater.* 45: 3383. b) Jia, H., Tao, X., and Wang, Y. (2016). *Adv. Electron. Mater.* 2: 1600136. c) Jia, H., Ju, Z., Tao, X. et al. (2017). *Langmuir* 33: 7600.
17 Zhao, D., Martinelli, A., Willfahrt, A. et al. (2019). *Nat. Commun.* 10: 1093.
18 Wang, H., Zhao, D., Khan, Z.U. et al. (2017). *Adv. Electron. Mater.* 3: 1700013.
19 Han, S., Alvi, N.U.H., Granlöf, L. et al. (2019). *Adv. Sci.* 6: 1802128.

6

Thermoelectric Properties of Carbon Nanomaterials/Polymer Composites

Yue Shu[1], *Zhenghong Xiong*[1], *Yang Liu*[1], *Yongli Zhou*[1], *Meng Li*[1], *Yujie Zheng*[2], *Shanshan Chen*[2], *and Kuan Sun*[1]

[1] *MOE Key Laboratory of Low-grade Energy Utilization Technologies and Systems, CQU-NUS Renewable Energy Materials & Devices Joint Laboratory, School of Energy & Power Engineering, Chongqing University, Chongqing, P. R. China*
[2] *School of Energy & Power Engineering, State Key Laboratory of Power Transmission Equipment & System Security and New Technology, Chongqing University, Chongqing, P. R. China*

Authors Yue Shu, Zhenghong Xiong, and Yang Liu are contributed equally to this work.

6.1 Introduction

Thermoelectric (TE) materials are considered to be a green energy material that enables direct conversion between thermal energy and electrical energy. Because there is no moving part or working fluid when incorporating thermoelectric materials in a device, the material can be widely used in thermoelectric generators, thermoelectric refrigeration devices, wearable electronics, and even aerospace applications [1]. The energy conversion efficiency of thermoelectric materials is evaluated by the dimensionless figure of merit (ZT) value, which is expressed as $ZT = S^2\sigma T/\kappa$, where S is the Seebeck coefficient (also known as thermopower, thermoelectric power, or thermoelectric sensitivity), σ is the electrical conductivity, T is the absolute temperature, and κ is the thermal conductivity. In most cases, these thermoelectric parameters are coupled to each other. Therefore, increasing the ZT value of TE materials has always been the focus of research in thermoelectric field. Power factor (PF) is often used to characterize thermoelectric properties when the thermal conductivity is constant or low. PF is represented by $S^2\sigma$. High ZT or PF values imply better thermoelectric performance. Conventional inorganic thermoelectric materials are mainly based

Flexible Thermoelectric Polymers and Systems, First Edition. Edited by Jianyong Ouyang.

on rare metal elements such as Bi, Pb, Sb, Ge, including Bi_2Te_3, SiGe, etc. [2–6]. However, these inorganic materials are expensive, toxic, hard to process, rigid, and brittle, which limit their use in flexible electronic devices. Recently, polymers and their composite materials have been receiving more and more attention due to their advantages of low cost, easy preparation, nontoxicity, and mechanical flexibility, which are suitable for various wearable applications.

Carbon materials contain hybridized SP, SP^2, and SP^3 bonds, thus they can form a rich and diverse carbon allotropes. Carbon-based thermoelectric materials are generally characterized by high flexibility, high strength, high temperature stability, good electrical properties but also high thermal conductivity [7]. So they are often added into polymer matrix to form thin film composite materials. In this way, the PF can be enhanced while the thermal conductivity of the composite material can be low. Typical filler carbon materials include carbon nanotube (CNT) [8–10], graphite [11, 12], graphene nanoplatelet (GNP) [13–15], reduced graphene oxide (rGO) [16, 17], etc. Though there are a number of excellent review articles covering the topics of organic thermoelectric materials [18–21], this review aims to focus only on the thermoelectric composite materials that contain carbon nanomaterials as the fillers in a polymer matrix. The polymers are classified into conducting polymers and non-conductive polymers. The materials properties, processing/treatment method, molecular interactions/structures, and thermoelectric performance will be introduced to highlight the structure–property relationship.

6.2 Conducting Polymers

Conducting polymers (CPs) have a typical π-conjugated backbone structure, i.e. alternating single and double bonds in the polymer backbone [22]. Conducting polymers are considered to be a new class of thermoelectric materials due to their low cost, nontoxicity, easy processability, mechanical flexibility, light weight, and intrinsically low thermal conductivity [11, 23]. The commonly investigated conducting polymers as TE materials are polyaniline (PANI) [11, 17, 24–29], polypyrrole (PPy) [30–36], poly(3,4-ethylenedioxythiophene) (PEDOT) [4, 37–39], poly(3-hexylthiophene) (P3HT) [40, 41], poly-Schiff bases (PSB) [42, 43], etc. Their chemical structures are displayed in Scheme 6.1. The addition of carbon materials as fillers into the polymer matrix can change the thermoelectric properties of these conducting polymers by introducing more interfaces and intermolecular interactions. The following session will review the carbon

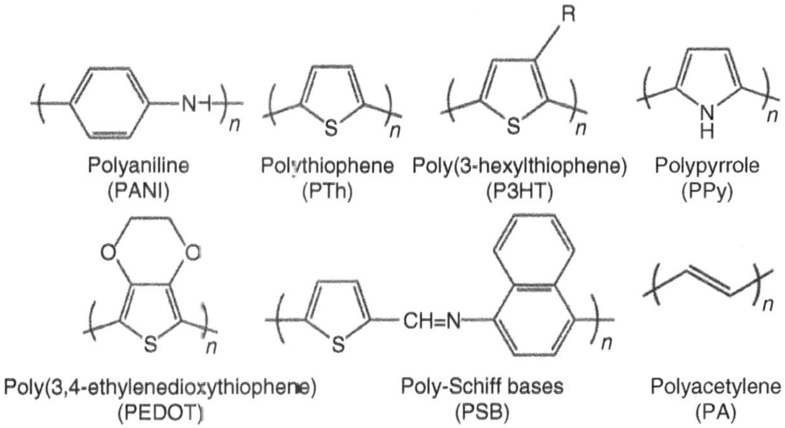

Scheme 6.1 Chemical structure of typical conjugated polymers in TE composites.

materials as fillers in PEDOT, PANI, PPy, and other conducting polymers in sequence.

6.2.1 PEDOT:PSS

Among all the conductive polymers, poly(3,4-ethylenedioxythiophene):poly(styrene sulfonate) (PEDOT:PSS) is the most popular one due to its tunable electrical conductivity, good transparency in visible range, excellent stability, and commercial availability [44]. Since 2008, studies on the thermoelectric properties of PEDOT:PSS and its composite have attracted great attention [44, 45]. In terms of the carbon material/PEDOT:PSS composites, most of the research studies focused on combining CNTs or GNPs with PEDOT:PSS to form the composite. Table 6.1 summarizes the TE properties of PEDOT:PSS-based composites combined with carbon nanomaterials.

6.2.1.1 CNT/PEDOT:PSS

In general, the thermoelectric parameters are strongly correlated to each other in bulk TE materials. For example, an increase in electrical conductivity often leads to an increase in thermal conductivity, making it difficult to enhance ZT value. Nevertheless, studies of CNTs indicate that it is possible to improve the electrical conductivity while keeping the thermal conductivity relatively constant [46]. This is because the individual CNTs are connected electrically at the junctions via π–π interaction, but phonons can hardly propagate across these junctions. Nevertheless, CNTs naturally form bundles due to van der Waals force among them. Without stabilizers/

Table 6.1 Summary of thermoelectric properties of carbon material/PEDOT:PSS composites.

Matrix	Filler	Method	Loading (wt%)	S (μVK^{-1})	σ (Scm^{-1})	PF (μWm^{-1} K^{-2})	κ (Wm^{-1}·K^{-1})	ZT	Reference
PEDOT:PSS	SWNT	Solution mixing	35%	25	400	25	0.4	0.02	[46]
PEDOT:PSS	SWNT	Solution mixing	60%	41	1350	160	0.2–0.4	/	[47]
PEDOT:PSS	SWNT	Two-step spin casting	/	38.9	241	21.1	/	/	[44]
PEDOT:PSS	DWNT	Post-treatment	20%	43.7	780	151	/	/	[48]
PEDOT:PSS	SWNT	Vacuum filtration	60%	44.3	600	105	0.26	0.12	[10]
PEDOT:PSS	SWNT	Wet-spinning and post-treatment	15%	14000	0.064	1200	60	0.7	[49]
PEDOT:PSS	GNP	Mixing and sonication	2%	58.77	32.13	11.09	0.14	0.021	[50]
PEDOT:PSS	rGO	Wet-chemical route	3%	17	1160	32.6	0.3	0.03	[51]
PEDOT:PSS	GNP	In situ polymerization	3%	26.8	637	45.7	/	/	[52]
PEDOT:PSS	GNP	Vacuum filtration	3%	22.0	1283.5	38.6	0.3	0.05	[53]
PEDOT	rGO	In situ polymerization	16%	31.8	50.8	5.2	/	/	[54]
PEDOT:PSS	CNT	Hot-pressing	50%	49.3	2350	572	/	/	[55]
PEDOT:PSS/ TCPP	DWNT	Dual-stabilizer	40%	70.0	960	500	0.14	/	[56]
PEDOT:PSS	MWNT	Rod-coating	2%	18.5	74	/	/	/	[57]
PEDOT:PSS	SWNT	Solution mixing	6.7%	59	1350	464	/	/	[58]
PEDOT:PSS	SWNT	Dilution-filtration	50%	30	900	83.9	/	/	[59]
PEDOT:PSS	SWNT	Liquid-phase exfoliation	95%	15	4000	90	0.53	0.03	[60]
PEDOT	SWNT	Chemical oxidation polymerization	15	17.5	570.4	19	/	/	[61]

surfactants, it is difficult to disperse highly entangled CNTs in organic solvents or water. In recent years, studies have shown that PEDOT:PSS can stabilize and disperse nanotubes effectively in water [46], thereby can assist in formation of CNT networks and prevent aggregation of CNT in thin films.

Kim et al. [46] first reported that the thermoelectric properties of CNT/polymer composites could be modified by PEDOT:PSS. They used dimethyl sulfoxide (DMSO)-doped PEDOT:PSS to disperse CNTs in water and compared the transport properties with an electrically insulating matrix, i.e. gum arabic (GA). The conductivity of PEDOT:PSS composites is much higher than that of GA composites, and increases with increasing CNT concentration in the polymer matrix. Drying at elevated temperatures also influences conductivity, because the polymer particles can push the CNTs into interstitial spaces to form a close contact during solvent evaporation, leading to a segregated and electrically connected network (Figure 6.1b). At high single-walled carbon nanotube (SWNT) concentration (35 wt%), conductivity of 400 $S\,cm^{-1}$ was recorded, 100 times higher than that of typical polymeric composites based on poly(3-octylthiophene) [62] or poly(methyl methacrylate) [63]. The increase in electrical properties did not significantly change the thermal conductivity, which could be maintained around 0.2–0.4 $W\,m^{-1}\cdot K^{-1}$. The best thermoelectric figure of merit (ZT) in this study was estimated to be 0.02 at room temperature. They attributed the good performance arose from the enhanced electrical connectivity between PEDOT:PSS and CNTs as well as among individual CNTs after the introduction of PEDOT:PSS. At the same time, thermal transport across the

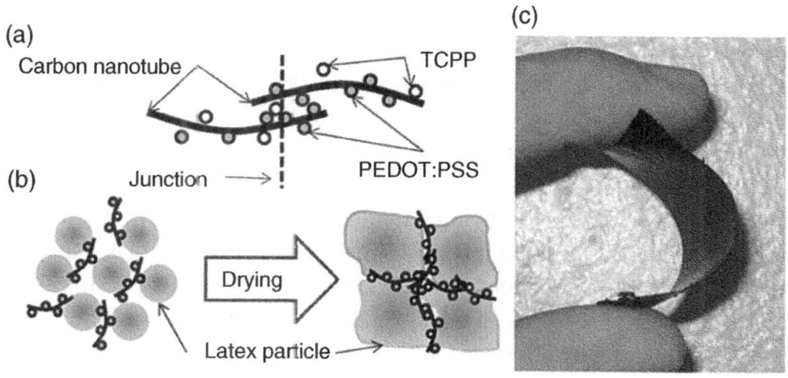

Figure 6.1 (a) Schematic of carbon nanotubes decorated by PEDOT:PSS and TCPP molecules and the junction formed between them. (b) Schematic depicting the formation of a segregated network composite during polymer coalescence as it dries. (c) Photo of a fully dried, free-standing, flexible composite. *Source:* Moriarty et al. [60]. © John Wiley & Sons.

tube–tube junctions is impeded due to mismatches in bond structures and vibrational spectra between CNT and PEDOT:PSS.

Moriarty et al. [60] studied the electrical properties and thermoelectric behavior of organic thin film nanocomposites, in which the fillers were prepared by liquid phase stripping. SWNTs were stabilized in aqueous solution using intrinsically conductive PEDOT:PSS. The conductivity σ increases linearly with the addition of SWNTs in the loading range between 20 and 95 wt%. At 95 wt%, these films exhibited metallic conductivity of $4000\,S\,cm^{-1}$. As σ increased, the S remained relatively constant, resulting in a maximum PF of $140\,\mu W\,m^{-1}\cdot K^{-2}$.

Aiming at better TE properties, Moriarty et al. [56] further prepared thermoelectric polymer nanocomposites with two different nanotubes, i.e. multi-walled carbon nanotube (MWNT) and double-walled carbon nanotube (DWNT), two CNT stabilizers, e.g. a conductive PEDOT:PSS and a semi-conductive meso-tetra(4-carboxyphenyl) porphine (TCPP), and four different CNT loadings in this dual-stabilizer system. First, CNT bundles are exfoliated and stabilized by PEDOT:PSS and TCPP in aqueous solution, as shown in Figure 6.1a. Then during drying, polymer particles aggregate, and force CNTs into the space between them (Figure 6.1b). The coalescing of polymer particles leads to the formation of a three-dimensional network of electrically connected but thermally disconnected nanotubes in flexible composites. Figure 6.1c displays such a free-standing and bendable composite thin film. With the increase of the concentration of MWNTs in PEDOT:PSS, the conductivity of these dual-stabilizer organic composites increases to about $95\,S\,cm^{-1}$. The Seebeck coefficient and thermal conductivity are still relatively unaffected by the increase of concentration, i.e. $40\,\mu V\,K^{-1}$ and $0.12\,W\,m^{-1}\cdot K^{-1}$, respectively. The replacement of MWNT with DWNT can increase σ and S to about $960\,S\,cm^{-1}$ and $70\,\mu V\,K^{-1}$ at 40 wt% DWNT loading, corresponding to a PF of approximately $500\,\mu W\,m^{-1}\cdot K^{-2}$ at room temperature. The high conductivity is ascribed to the stabilization of CNTs by PEDOT:PSS and TCPP possibly through π–π stacking, that bridges the individual nanotubes to facilitate efficient charge carrier transport in these composites. The thermal conductivity is relatively unaffected by the component ratio between the two stabilizers, the CNT concentration and its type. This may be due to the mismatch of the vibrational spectra at many interfaces in the composite, which hinders the phonon transport.

Song et al. [44] fabricated a bilayered SWNT/PEDOT:PSS composite by a two-step spin-casting method. In a typical process, the hydrophilic glass surface is first coated with SWNTs and dried at 60 °C for 12 hours, then a PEDOT:PSS layer (or doped with 5 vol% DMSO) is deposited on top to form the two-layered SWNT/PEDOT:PSS composite film. The maximum conductivity and Seebeck coefficient of the composites are 2.41 S cm^{-1} and 38.9 μV K^{-1}, respectively. The calculated PF can reach 21.1 μW $m^{-1}{\cdot}K^{-2}$. Scanning electron microscopy (SEM), Fourier transform infrared spectroscopy (FTIR), and Raman spectroscopy are employed to study the influence of the CNTs characteristics on the morphological, electrical, and thermoelectric properties of the composite material. It is speculated that the addition of SWNTs weakens the interaction between the PEDOT and insulating PSS chains, thus the electrical conductivity increases.

The TE properties of CNT/PEDOT:PSS composites can be improved by post-treatment. Lee et al. [48] carried out ethylene glycol (EG) treatment by immersing the CNT/PEDOT:PSS composite film in EG for 1 hour, followed by thermal annealing at 140 °C for 10 minutes. The EG treatment can effectively remove excess PSS chains that are not bonded with PEDOT via Coulomb force, thus greatly enhancing the conductivity of the PEDOT:PSS matrix. The EG-treated composite film containing 20 wt% CNTs exhibits a maximum PF of 151 μW $m^{-1}{\cdot}K^{-2}$. The results demonstrated that EG treatment was an effective method to prepare cost-effective, high performance thermoelectric CNT/PEDOT:PSS composite films.

Hsu et al. [58] reported the treatment of CNT/PEDOT:PSS films of different CNT loadings with DMSO or formic acid (FA). While the electrical conductivity was monotonically increased, the thermopower of the pristine PEDOT:PSS (36 mV K^{-1}) was raised with the addition of CNT up to 59 mV K^{-1} at 6.7 wt% CNT and then decreased with more CNT to 50 mV K^{-1} at 20 wt% CNT. Post-treatment could significantly enhance σ while S was only slightly reduced, leading to a maximum PF of 464 and 407 μW $m^{-1}{\cdot}K^{-2}$ after DMSO and FA treatment, respectively. Hall measurement suggested the DMSO/FA treatment increased both mobility and carrier concentration of pristine PEDOT:PSS. The removal of PSS raised the relative weight ratio of conducting PEDOT (i.e. increased carrier concentration), as well as led to more efficient carrier transport through the conducting PEDOT due to less hindrance by insulating PSS (i.e. increased mobility). In addition, it is necessary to optimize CNT loading to fine-tune the TE properties. In this study, a CNT loading of 6.7 wt% showed the highest S and PF. The CNT loading was much lower than other typical CNT/PEDOT:PSS composites reported. The authors claimed a moderate CNT percolation could maximize the energy filtering effect at the CNT/PEDOT:PSS junction.

Thermoelectric properties of composites are largely dependent on processing methods. Lee et al. [55] proposed a post-processing method to improve the thermoelectric PF of CNT/conjugated polymer nanocomposite films. Figure 6.2 is a schematic diagram of the preparation process of

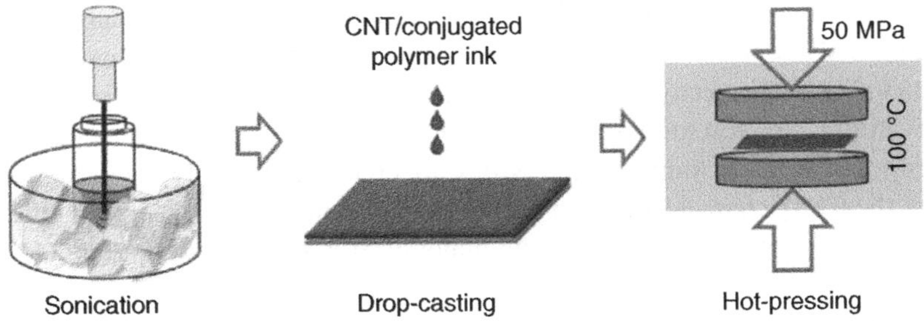

Figure 6.2 Schematic diagram of the fabrication process of CNT/conjugated polymer nanocomposite films. *Source:* Lee et al. [55]. © Elsevier.

CNT/conjugated polymer nanocomposite films. In such a process, few-walled CNT (FWNT, 50 wt%) were first mixed with either P3HT (50 wt%) in o-dichlorobenzene (oDCB) or PEDOT:PSS (50 wt%) in distilled deionized water with the aid of ultrasonication. Then, mixture was drop cast on glass. Finally, the nanocomposite films were densified by uniaxial hot pressing. It was found that the PF of the CNT/P3HT and the CNT/PEDOT:PSS nanocomposite film increased from 150 ± 9 to $217 \pm 30\,\mu W\,m^{-1}{\cdot}K^{-2}$ and from 371 ± 44 to $572 \pm 92\,\mu W\,m^{-1}{\cdot}K^{-2}$, respectively after hot pressing. An organic thermoelectric generator (OTEG) was fabricated, and the maximum output power of hot pressed CNT/P3HT OTEG and CNT/PEDOT:PSS OTEG was 46.0 and 36.3 nW, respectively, whereas the temperature difference was 10 K.

Jiang et al. [64] designed and prepared SWNT/PEDOT:PSS hybrid organic film by vacuum filtration. Figure 6.3 illustrates the film preparation process as well as the appearance of the respective solutions and vacuum filtrated thin films. At 60 wt% SWNT loading, the Seebeck coefficient of the composite film reaches $44.3\,\mu V\,K^{-1}$, which is more than three times of the pristine PEDOT:PSS ($14.1\,\mu V\,K^{-1}$). The optimal PF of the composite film is $105\,\mu W\,m^{-1}{\cdot}K^{-2}$ and the optimal ZT value is 0.12. In addition, the composite film can maintain a low thermal conductivity of the polymer sample and is insensitive to the loading of SWNTs. When the loading increases from 0 to 90 wt%, the thermal conductivity of composite film increases only slightly from 0.15 to $0.36\,W\,m^{-1}{\cdot}K^{-1}$. This low thermal conductivity is four orders of magnitude smaller than that of neat SWNT film, which signifies the importance of polymeric composite in maintaining a low thermal conductivity.

Choi et al. [49] found that ionic conductors could generate a large amount of thermally induced voltage (about $1\text{–}10\,mV\,K^{-1}$). Yet, the electrical conductivity of typical ionic conductors is several orders of magnitude lower than conventional TE materials. To compensate the electrical conductivity,

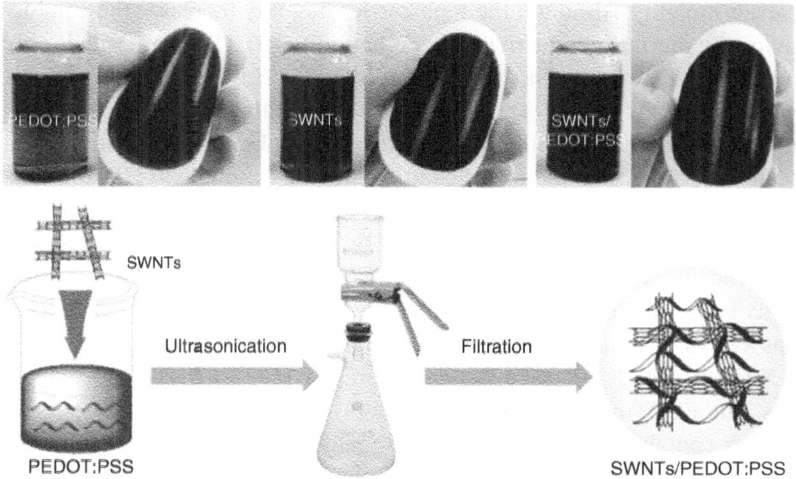

Figure 6.3 Schematic illustration of vacuum filtration process to fabricate the SWNT/PEDOT:PSS hybrid organic film. *Source:* Chen et al. [10]. © Royal Society of Chemistry.

they assembled the ion conductor PEDOT-Tos and CNTs into all-organic composites, and treated the composites by chemical reduction with tetrakis(dimethylamino)ethylene (TDAE) to obtain a thermopower of ~14 $mV\,K^{-1}$. Due to the presence of CNTs, the conductivity was maintained at a decent level, resulting in a PF of ~1200 $\mu W\,m^{-1}\cdot K^{-2}$ and ZT of 0.7 at room temperature. The incorporation of CNTs into PEDOT greatly enhanced the carrier mobility, which promotes the enhancement of TE properties. On the one hand, the electron contribution is large when the CNT concentration is high. On the other hand, when the CNT concentration is low or the reduction time is long, the ion contribution is large.

6.2.1.2 Graphene/PEDOT:PSS

Graphene is a typical two-dimensional crystalline carbon material that consists of a single atomic layer of carbon atoms arranged in honeycomb-like patterns. The conjugation in in-plane direction and π–π interaction between two graphene layers provide this material with high conductivity and good thermoelectric properties [4]. Similar to CNT, graphene is a suitable candidate filler material to form a composite.

Kim et al. [50] prepared high-performance thermoelectric graphene/PEDOT:PSS films by solution spin-coating. The graphene content is less than 3 wt%, which is enough to facilitate efficient carrier transport by interacting with PEDOT:PSS via strong π–π interaction, thereby increasing carrier mobility. Although the high carrier mobility increases the conductivity, the thermal conductivity of graphene/PEDOT:PSS composite is only $0.14\,W\,m^{-1}\cdot K^{-1}$, lower than that of the PEDOT:PSS film containing 35 wt% of SWNT ($0.4\,W\,m^{-1}\cdot K^{-1}$). This behavior can be attributed to two mechanisms: (i) the porous multilayer structure of graphene acts as the phonon scattering center that reduces the lattice thermal conductivity; (ii) the electronic thermal conductivity (k_e) that is proportional to the electrical conductivity through the Wiedemann–Franz law, $k_e = ne\mu LT$, is lower than the influence of lattice thermal conductivity (k_l) because the thermal conductivity is dominated by phonons. The PEDOT:PSS film with 2 wt% graphene has a PF of $11.09\,\mu W\,m^{-1}\cdot K^{-2}$ at 300 K and a maximum ZT of 0.021.

Xu et al. [54] developed a strategy to prepare rGO/PEDOT nanocomposites. The EDOT monomer was added stably into the rGO water dispersion system containing sodium polystyrene sulfonate (PSSNa). As shown in Figure 6.4, under mechanical agitation, i.e. ultrasonic treatment and stirring, EDOT molecules adsorbed on the surface of rGO nanosheets due to strong π–π interaction. Subsequently, template-directed in situ polymerization took place on the surface of rGO, and then the rGO nanosheets were coated by thick PEDOT layers on both sides of the rGO nanosheets. Thus, the derivative structures with rGO nanosheets and PEDOT layers as the core and shell layers were constructed. Compared with neat PEDOT, rGO/PEDOT nanocomposites exhibited significant improvement in TE performance at room temperature, with a PF of $5.2\pm0.9\,\mu W\,m^{-1}\cdot K^{-2}$, which is 13.3 times of that of PEDOT.

Yoo et al. [52] used PSS as dispersant of graphene and dopant of PEDOT. After effective dispersion, content optimization and in situ polymerization with PSS, graphene/PEDOT:PSS composites with high conductivity were synthesized directly on a liter scale. After introducing 3 wt% graphene, the conductivity of the films is $637\,S\,cm^{-1}$ and PF reaches $45.7\,\mu W\,m^{-1}\cdot K^{-2}$, which is 93% higher than that of pristine PEDOT:PSS. The advantages of this method include (i) graphene exhibits better dispersion in

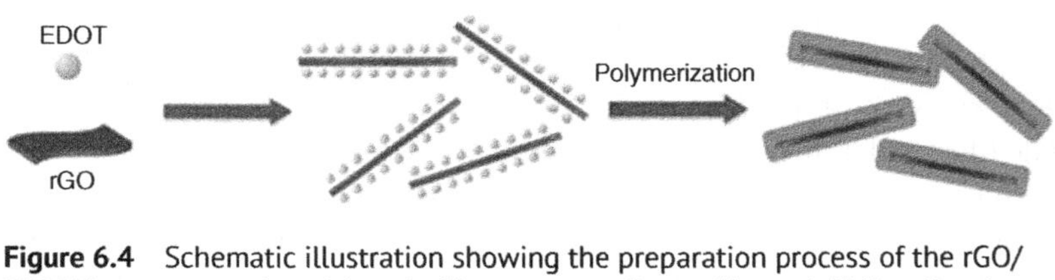

Figure 6.4 Schematic illustration showing the preparation process of the rGO/PEDOT nanocomposite with a pie-like structure. *Source:* Xu et al. [54]. © Royal Society of Chemistry.

aqueous solution when using high molecular weight polyelectrolytes, e.g. PSS, rather than small molecule-based dispersant; (ii) a large quantity of graphene can be used in large-scale production of composite materials.

Li et al. [51] prepared a GO/PEDOT:PSS film by solution means and selected HI as a reductant to prepare rGO/PEDOT:PSS composite films via a simple wet chemical route. The in-plane conductivity and the Seebeck coefficient of the films were measured at room temperature. For 3 wt% GO/PEDOT:PSS films after HI reduction, the maximum PF is ~32.6 $\mu W m^{-1} \cdot K^{-2}$, which is about 1.5 times that of pure PEDOT:PSS films prepared by the same method. The films were characterized by X-ray diffraction (XRD), FTIR, X-ray photoelectron spectroscopy (XPS), and SEM. It is concluded that the reason for the improvement is the interaction between rGO and polymer matrix and the removal of some PSS from the film. This interaction provides a channel for carrier transport and makes the packaging of PEDOT:PSS chains more orderly.

In recent years, studies on the thermoelectric properties of graphene/PEDOT:PSS have focused on the combination of rGO and PEDOT:PSS. However, due to the structural defects of rGO, the current research has not sufficiently synergized the advantages of PEDOT:PSS and graphene. The exfoliated graphene has fewer structural defects and higher conductivity than rGO. Xiong et al. [53] successfully prepared flexible graphene/PEDOT:PSS composite nano-films with liquid exfoliated graphene and PEDOT:PSS solution by direct vacuum filtration. The nano-film has a high conductivity (1283.5 $S cm^{-1}$) and enhanced Seebeck coefficient (22.0 $\mu V K^{-1}$). In the case of 3 wt% graphene, the TE power factor of the nano-film is 38.6 $\mu W m^{-1} \cdot K^{-2}$. The most important contribution of graphene is to increase the Seebeck coefficient of the initial PEDOT:PSS. This is because the conjugated electronic system of graphene can interact with PEDOT:PSS to provide a channel for carrier transport. In addition, the authors found that hydrazine (HZ) treatment had a positive effect on improving the Seebeck coefficient. By adjusting the concentration of hydrazine, treatment time and temperature, the optimum PF of the nano-films reached 53.3 $\mu W m^{-1} \cdot K^{-2}$. Assuming that the thermal conductivity in the through plane direction is 0.3 $W m^{-1} \cdot K^{-1}$, the corresponding ZT is estimated to be 0.05 at room temperature, which is higher than previous reports on rGO/PEDOT:PSS films.

6.2.2 Polyaniline (PANI)

PANI is a conducting polymer widely used as a thermoelectric material. Due to relatively low crystallinity, a typical PANI film consists of conductive islands scattered in disordered microstructures. By adding carbon nanomaterials into the PANI matrix, the TE properties can be improved by one or more effects, such as reducing disorder of PANI, tunable carrier concentration, formation of more interfaces, and establishment of a good conductive network. Since all these effects are largely related to processing techniques and conditions, the following session will focus on different methods in obtaining the carbon nanomaterials/PANI composite films, whose TE properties are summarized in Table 6.2 [72].

6.2.2.1 Powder Mixing Method

Powder mixing method is a simple process to fabricate carbon nanomaterial/PANI composites. First, the carbon nanomaterials and conducting polymers are prepared separately. Then these two materials in powder form are mechanically ground and mixed, and the composite material is obtained by tableting or annealing. Without the aid of dispersant, Zhang et al. [68] mixed MWNT and PANI at −196 °C using liquid nitrogen, followed by consolidating the obtained powder by spark plasma sintering (SPS) at 100 °C, in which the powder sample was thermally annealed under an electric field so that Joule heating within the sample could achieve near theoretical density at a low sintering temperature. When the MWNT loading increased from 10 to 30 wt%, the electrical conductivity of the PANI/MWNT composite was greatly enhanced from 0.0351 to 0.159 $S\,cm^{-1}$. At the same time, the Seebeck coefficient kept constant at around 25 ± 5 $\mu V\,K^{-1}$. So the optimal PF was 0.1073 $\mu W\,m^{-1}{\cdot}K^{-2}$. The increase in conductivity is in accordance with the known percolation law, which predicts that when the filling volume of conductive particles reaches the percolation threshold, the conductivity will increase significantly [17].

Abad et al. [14] added GNPs with different mass ratios as conductive filler to PANI, and measured the thermoelectric parameters of the GNP/PANI composites. The GNP/PANI composite powder was prepared by mechanically grinding GNP powder and PANI powder with a glass mortar and a pestle. The PANI powder was obtained by oxidative polymerization in HCl solution. In order to facilitate the TE measurement, the PANI/GNP powder was cold pressed into solid pellets. The SEM results (Figure 6.5) showed that the GNPs were mostly parallel to the pellet surface in the PANI matrix. In addition, the GNP clusters were developed as the GNP content increased. The density of pure PANI and pure GNP, as well as the mass fraction of each component were measured to obtain a volume fraction of GNP. The thermoelectric properties were then measured as a function of GNP volume. The variation of electrical conductivity was a function of the GNP volume fraction, which was found to be consistent with the percolation law in

Table 6.2 Summary of thermoelectric properties of PANI-based composite films.

Matrix	Filler	Synthesis method	Loading (wt%)	S (μVK^{-1})	σ (Scm^{-1})	PF (μWm^{-1} K^{-2})	κ (Wm^{-1}·K^{-1})	ZT	Reference
PANI	MWNT	In situ chemical polymerization	44%	23.3	40.35	/	0.29	0.0022	[23]
PANI	GNP	In situ chemical polymerization	57%	33	59	/	15	0.00015	[13]
PANI	GNP	Solid-state grinding and drop coasting	45%	15	850	19	/	/	[65]
PANI	MWNT	In situ chemical polymerization	93%	79.8	14.1	/	0.27	0.01	[66]
PANI	S/DWNT	Solution mixing	50%	119	29.95	42	/	/	[67]
PANI	MWNT	Cryogenic grinding	30%	26	1.59	0.1073	0.79	0.000076	[68]
PANI	GNP	Mechanically grinding	50%	33	123	14	/	/	[14]
PANI	SWNT	Solution mixing	64%	47.8	769	176	0.43	0.12	[9]
PANI	DWNT	Solution mixing	30%	61	610	220	/	/	[69]
PANI	CNT	In situ chemical polymerization	40%	10	17.1	0.17	/	/	[8]
PANI	GNP	In situ chemical polymerization	30%	28.7	29	2.4	/	/	[70]
PANI	GNP	In situ chemical polymerization	3%	14.5	604	12.7	0.22	0.74	[29]
PANI	CNT	In situ chemical polymerization	94%	37	2012	273	/	/	[28]
PANI	DWNT&GNP	Layer-by-layer	/	130	1080	1825	/	/	[71]

Source: Based on Liu et al. [72]. © John Wiley & Sons.

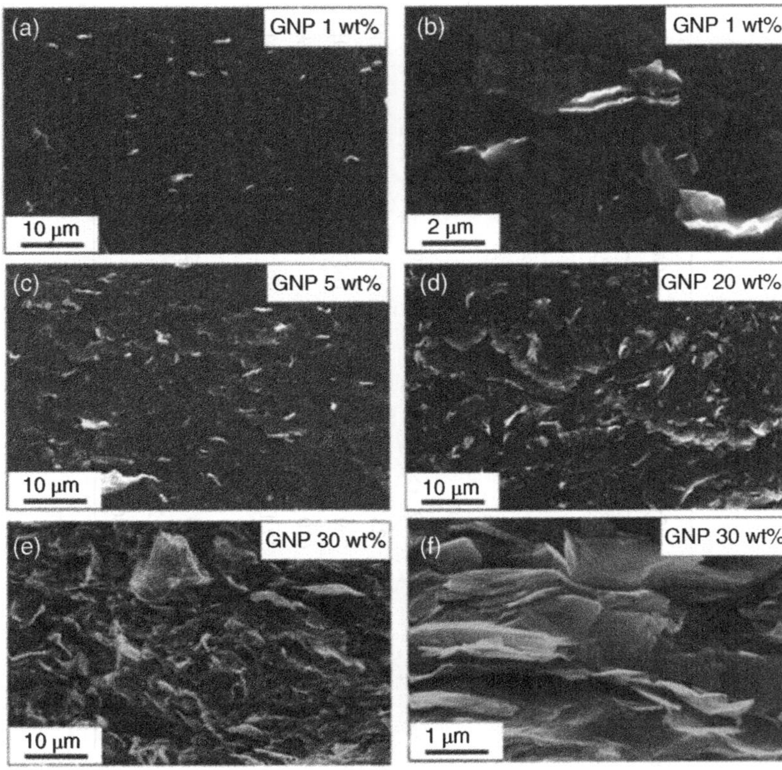

Figure 6.5 FE-SEM images of the fracture perpendicular to the pressure direction of the PANI/GNP pellets of (a) and (b) GNP 1 wt%, (c) GNP 5 wt%, (d) GNP 20 wt%, and (e) and (f) GNP 30 wt%. *Source:* Abad et al. [14]. © Royal Society of Chemistry.

composites. The Seebeck coefficient increased up to 33 $\mu V K^{-1}$ while the original Seebeck coefficient for PANI and GNP were 2.6 and 17.3 $\mu V K^{-1}$, respectively. The increase in the Seebeck coefficient of PANI/GNP composites was attributed to the energy filtering effect provided by the boundaries between GNP and PANI, which was able to scatter low energy carriers and thus improving the average charge carrier energy. The synergistic increase in conductivity and Seebeck coefficient delivered a maximum PF of 14 $\mu W m^{-1} \cdot K^{-2}$ at an optimal GNP loading of 50 wt%. The percolation threshold is related to the dispersion of the fillers in the polymer matrix. More carbon nanomaterials means clusters are more likely to occur. Thus, better dispersion are favored, so that lower percolation thresholds can be achieved.

6.2.2.2 Solution Mixing Method

In solution mixing method, the carbon nanomaterials and the conducting polymer are dispersed in the same solution. Ultrasonication or stirring are often employed to promote thorough mixing and dispersion of the two. Good dispersion reduces the chances of agglomeration of carbon nanomaterials, thus helping to form a conductive network. For example, by using this method Yao et al. [9] prepared CNT/PANI TE composite material. It was observed under SEM that the CNTs were well dispersed in the PANI matrix under low CNT loading of 30 wt%. As the CNT content increases, CNTs appear more and more exposed and aggregated. Uniform dispersion and solution processes provide greater contact area for carbon nanoparticles and PANI matrices and enhance their interaction. Although the lower load has better dispersion, the addition of more conductive CNTs increases the conductivity. After balancing the amount and dispersion of conductive particles, they obtained an optimum conductivity of 749 $S\,cm^{-1}$ and a PF of 176 $\mu W\,m^{-1}\cdot K^{-2}$ at a load of 64 wt%.

Wang et al. [65] compared the thermoelectric properties of graphene/camphorsulfonic acid (CSA)-doped PANI composites obtained by direct powder mixing and solution-assisted dispersion preparation. For the solution-assisted dispersion preparation, the graphene and PANI powder after mixed and ground were dispersed in m-cresol solvent aided by ultrasonication at room temperature. Then the mixed solution was drop cast on glass to obtain the TE sample. It was found that the material prepared by powder mixing had a lower electrical conductivity and PF than the material dispersed in a solution. In the m-cresol solution, the conformation of the PANI molecular chain can change from a compacted coil to an expanded coil, as demonstrated in Figure 6.6. More stretched PANI molecular chain

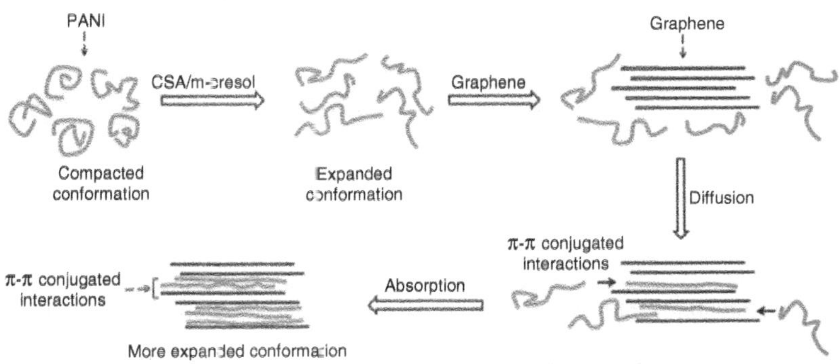

Figure 6.6 Schematic illustration of the synthesis procedure of PANI/graphene composite film by a simple solution-assistant dispersing process. *Source:* Wang et al. [65]. © Royal Society of Chemistry.

structure enhances the delocalization of polarons and bipolarons along the polymer chain, resulting in greater carrier mobility. Thus they obtained the optimal PF of 19 $\mu W m^{-1} \cdot K^{-2}$ under 27 wt% graphene loading, while the direct powder mixed sample only had a PF of 8 $\mu W m^{-1} \cdot K^{-2}$ under the same loading.

The increase of the Seebeck coefficient in CNT/PANI composite is related to the interfacial effect between PANI and carbon nanomaterials. Wang et al. [67] prepared un-doped PANI and CSA-doped PANI (PANI-CSA) in advance, and then mixed with CNTs in a ratio of one to one. To estimate the height of energy barrier at the interface, they measured the highest molecular orbital (HOMO), the lowest unoccupied molecular orbital (LUMO), and the Fermi level. Figure 6.7 schematically shows the energy barrier and carrier transport at the interface. The transport of carriers with energy less than the energy barrier can be hindered, which increases the average energy of the carriers. The PANI shows higher energy barrier than PANI-CSA, thus resulted in higher Seebeck coefficient. However, the conductivity of the un-doped PANI was too low, so they employed HCl doping process to improve the conductivity of CNT/PANI, and obtained a conductivity of 29.95 $S cm^{-1}$

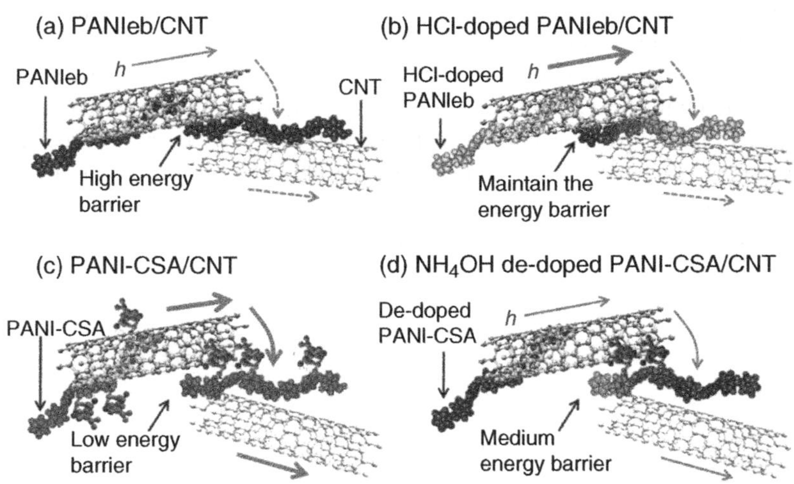

Figure 6.7 Proposed hole carrier transport at CNT-PANI-CNT junctions. (a) The resistive junction has a high energy barrier, reducing the number of carriers passing through the interface. (b) The overall carrier concentration increases due to the HCl doping of PANI, while the PANI at the junction is less affected due to $\pi-\pi$ interactions, maintaining the energy barrier. (c) Electrically conducting PANI-CSA creates a low energy barrier. (d) The de-doping decreases the overall carrier concentration and affects the PANI-CSA between CNTs due to the weaker $\pi-\pi$ interaction. *Source:* Wang et al. [67]. © Elsevier.

and a Seebeck coefficient of 119 $\mu V K^{-1}$, and the optimal PF of 42 $\mu W m^{-1} \cdot K^{-2}$ in CNT/HCl-PANI composites.

Previous studies have shown that the Seebeck coefficient and electrical conductivity are inversely correlated with increasing carrier concentration. Carbon nanomaterials filled in PANI matrix decoupled thermoelectric parameters to a certain extent, thus achieving a synergistic enhancement of the Seebeck coefficient and conductivity. Wang et al. [69] prepared DWNT/PANI and DWNT/PANI-CSA composites in m-cresol solution. Hall measurements were used to measure carrier concentration and mobility of DWNT/PANI-CSA composites. The results showed that with the addition of DWNT, the carrier mobility increased by 50 times (from ~0.15 to ~7.3 $cm^2 V^{-1} \cdot s^{-1}$), while the carrier concentration was only reduced by four times (from ~2.1×10^{21} to ~5.6×10^{20} cm^{-3}). Although the decrease in carrier concentration increases the Seebeck coefficient and decreases the conductivity, the large increase in carrier mobility compensates for the conductivity, resulting in a synergistic increase in the Seebeck coefficient and conductivity. In addition, the improved carrier mobility could attribute to the band alignment of CNT and PANI-CSA, which aid hole carrier transport in the CNTs with higher carrier mobility. Therefore, the maximum PF they obtained was ~220 $\mu W m^{-1} K^{-2}$ at 30 wt% DWNT.

6.2.2.3 In Situ Polymerization Method

In addition to the abovementioned two methods, in situ polymerization using carbon nanomaterials as a template has been exploited. In such a method, carbon filler and the polymer monomer are dispersed in a solvent, and then polymerization of the monomer takes place to form a composite material.

Wang et al. [8] compared the thermoelectric properties of PANI/MWNT materials prepared using in situ polymerization and mixing in HCl solution. The in situ polymerization method employed MWNT as a template to guide the polyaniline molecules to extend orientation along the surface of the one-dimensional CNT. The XRD, Raman, and SEM techniques were used to confirm the more ordered arrangement of PANI in MWNT/PANI prepared using in situ polymerization. The strong π–π interaction between CNT and PANI resulted in the formation of a core–shell structure, with PANI polymer chains wrapped around CNTs. When the CNT loading is 40 wt%, the conductivity and the Seebeck coefficient of the composite film prepared by in-situ polymerization is 17.1 $S cm^{-1}$ and 10 $\mu V K^{-1}$, respectively. Similarly, Lu et al. [70] in situ polymerized PANI doped with HCl using conductive GNPs as templates. The resulting GNPs/PANI composite exhibited a ZT value of 1.95×10^{-3} at 30 wt% GN loading, with σ of ~29 $S cm^{-1}$, S of ~28.7 $\mu V K^{-1}$ and PF of ~2.4×10^{-6} $W m^{-1} \cdot K^{-2}$.

Using the same in situ polymerization method, a few studies aim to optimize thermoelectric performance via better dispersion of carbon fillers. The pristine graphene has few polar functional groups, so it exhibits poor dispersibility in polar solvents [29]. To improve it, Lin et al. [29] tried to copolymerize *p*-phenediamino-modified graphene (PDG) with aniline to form a semi-interpenetrating networks. Due to the good dispersion of the modification, the amount of graphene required was reduced drastically. The temperature at which graphene was functionalized and the amount of graphene added affected the properties of the composite. The surface of the modified graphene has an aryl group and is bonded by a covalent bond, which allows electrons to be transferred from the conjugated π-bond of the graphene to the covalent bond, which could increase the carrier concentration [73]. Though that decreased the Seebeck coefficient, the more optimized conductivity still improves the thermoelectric performance, obtaining a ZT of 0.74 at the modification temperature of 30 °C and the graphene content of 3 wt%.

Similarly to improve the dispersion, Chen et al. [23] prepared a 3D interconnected network of CNTs by chemical vapor deposition (CVD) in order to avoid uneven dispersion. The 3D CNT network was further doped with K or I_2 by gas fuming method. In situ polymerization of PANI on CNT network was carried out by adding ammonium peroxidisulfate (APS) as oxidant in HCl and ethanol mixed solvent. At a content of 44 wt% CNT and at 373 K, they obtained a composite with a ZT of 0.22×10^{-3}.

Since in situ polymerized PANI chains are oriented along the template, their thermoelectric properties can exhibit anisotropic behavior. To confirm this, Chen et al. [26] prepared a CNT template in a specific direction by CVD method and polymerized PANI on it (Figure 6.8). Conductivity parallel to the direction of CNT alignment was five times larger than in vertical direction. In addition, thermal conductivity exhibited a 25 times difference.

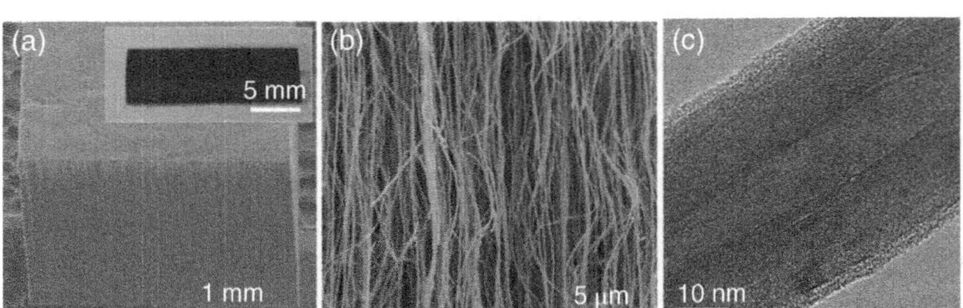

Figure 6.8 (a) Macro morphologies of the CNT arrays grown by CVD; (b) SEM morphology of the CNT arrays; (c) TEM morphology of the CNTs within the array. (A color version of this figure can be viewed online. *Source:* Chen et al. [26]. © Elsevier.

Hall measurements showed no significant difference in carrier concentration in the direction of parallel and vertical CNT alignment. The conductivity in the direction parallel to the CNT alignment is larger due to the larger carrier mobility. In 2018, Chen et al. [10] again prepared in situ polymerized SWNT/PANI and MWNT/PANI composite powder and cold pressed into a block. The cold pressing operation makes the PANI chain more inclined to align perpendicular to the cold pressing axis, which caused significantly different conductivity and thermal conductivity in parallel and perpendicular to the cold-pressed axis. When the CNT is 30 wt%, the maximum difference between the electrical conductivity and the thermal conductivity in the direction of the vertical and parallel cold pressing axes is reached, and the ratio reaches the magnitude up to 1.4. Finally, they obtained the optimal PF value at around 0.2 $\mu W cm^{-1} \cdot K^{-2}$ when SWNT is ~40 wt% in a direction parallel to the cold press axis.

The π–π interaction between carbon filler and PANI is a weak interaction, the conductive layer can be formed only in the inner layer close to carbon. Li et al. [28] used an amine-functionalized CNT (A-CNT) as a template to polymerize PANI in situ to form a composite film. As the A-CNT content in the film increases, the thickness of the shell PANI becomes thinner. However, only the inner PANI is more stable and remains on the CNT because of the weak π–π interaction. They soaked different amounts of CNT materials in dimethylformamide (DMF) solution and found that the A-CNT content was 14, 57 and 94 wt%, respectively. The PANI of the outer layer is weakly bound to the CNT and dissolved in the DMF, so the thicker the PANI shell, the deeper the blue color of the DMF solution. PANI that failed to dissolve in DMF formed a highly conductive layer due to π–π interaction, so they got the optimal PF of 273 $\mu W m^{-1} \cdot K^{-2}$ under 94 wt% A-CNT loading.

In addition to the dispersion of carbon fillers and the interfacial interaction with PANI, the doping state of PANI is also one of the factors affecting thermoelectric properties. Xiang and Drzal [13] explored the effect of PANI protonation level on electrical conductivity, thermal conductivity, and the Seebeck coefficient. They obtained in-situ polymerized GNP/PANI films on PVDF substrates by vacuum self-assembly technique. The prepared GNP/PANI materials were doped and de-doped in HCl or NH_4OH solution for two hours. Imine nitrogen (–N=) is preferentially protonated in GNP/PANI film by hydrochloric acid. They estimated the amount of Cl anion by XPS measurements and quantify the protonation level by the ratio of Cl anion to total N atom ($[Cl^-]/[N]$). Theoretically, when all the imine nitrogens are oxidized, the maximum protonation is 0.5. They found that the GNP/PANI material polymerized with 0.05 M monomer had a maximum Seebeck coefficient of 33 μV K^{-1} at a protonation level of 0.15–0.2. Further increase or decrease in protonation levels will cause a decline in the Seebeck coefficient. The introduction of GNP to some extent compensates for the lower conductivity caused by insufficient protonation. However, changes in thermal conductivity indicate that it is not directly affected by the level of protonation. Finally, the GNP/PANI prepared by in situ polymerization directly in the HCl solution has the largest ZT value of 1.51×10^{-4}.

Reducing the thermal conductivity is beneficial to improve the performance of the TE material and to maintain the temperature difference across the TE device. Phonon scattering is an effective method for reducing thermal conductivity in TE materials. Xiang and Drzal [15] prepared a graphene paper and polymerized PANI on the paper. The obtained 2D composite paper forms a 3D composite material by folding and bending (Figure 6.9). At the folded edge, due to the tensile and compressive stress, a structure in which the outer edge is broken and the inner edge is linked can be formed. This partially damaged structure hinders the propagation of phonons, but the paper is still conductive due to the tunneling effect of electrons. Further compaction of the folded GNP/PANI paper revealed an increase in the Seebeck coefficient and an increase in conductivity due to the addition of

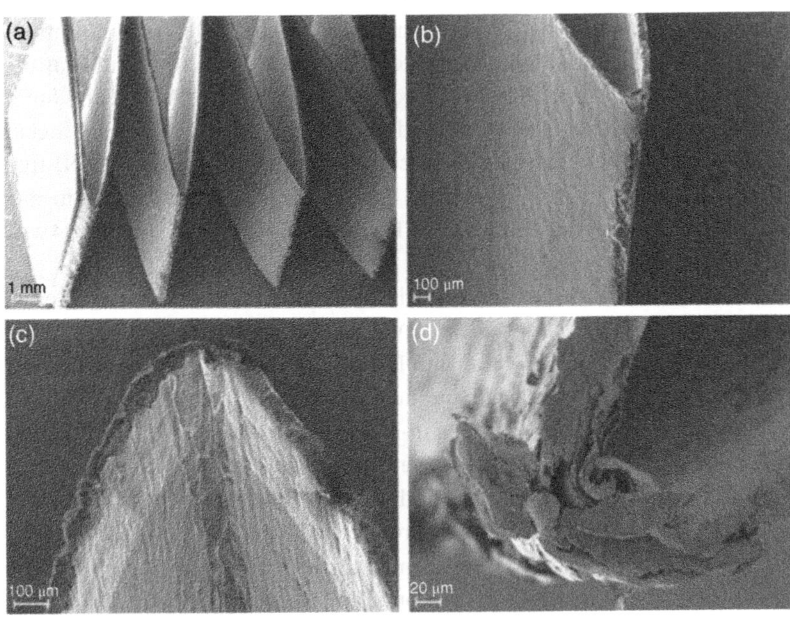

Figure 6.9 (a) SEM micrographs of a folded strip at low magnification, (b) the convex side of the folding edge, (c) the concave side of the folding edge, and (d) a magnified view of the tip of the folding edge. *Source:* Xiang and Drzal [15]. © Elsevier.

more transport paths. Eventually they obtained an unfolded paper with a ZT value of 3.7×10^{-6}, while that of the folded paper was 1.3×10^{-5}.

6.2.2.4 Layer-by-Layer (LBL) Deposition

The above methods all use the carbon nanomaterials as a dispersed phase to construct composite materials, and manipulate the thermoelectric properties through a two-phase interface and π–π interaction. Layer-by-layer (LBL) deposition is a preparation method that is employed in this field after 2010. It forms an ordered structure by intercalating carbon nanomaterial and polymer layers.

Cho et al. [71] used LBL deposition to deposit PANi/graphene/PANi/DWNT multilayers on a Si-wafer, and deposited such multilayers in cycles. The preparation procedure is represented in Figure 6.10a. Figure 6.10b explains the work mechanism. The PANI covers the DWNT because of the π–π interaction, which reduces the distortion and defects of the PANI chains and improves the carrier mobility. DWNT wrapped with PANI acts as a conductive bridge connecting the dispersed graphene, thus forming a conductive network to increase the carrier transport channels. The formation of a conjugate network synergistically enhances the Seebeck coefficient and conductivity. Finally, they obtained an optimal PF value of $1825\,\mu W\,m^{-1}\cdot K^{-2}$ at 40 cycles.

6.2.3 Polypyrrole (PPy)

PPy is another popular thermoelectric polymer because of its low cost, easy processing, and environmental stability. Combining PPy with carbon nanomaterials to form a composite material can improve thermoelectric parameters such as electrical conductivity and thermal conductivity over the pristine PPy film. Thermoelectric properties of TE composites using PPy or other conjugated polymers as the matrix are summarized in Table 6.3.

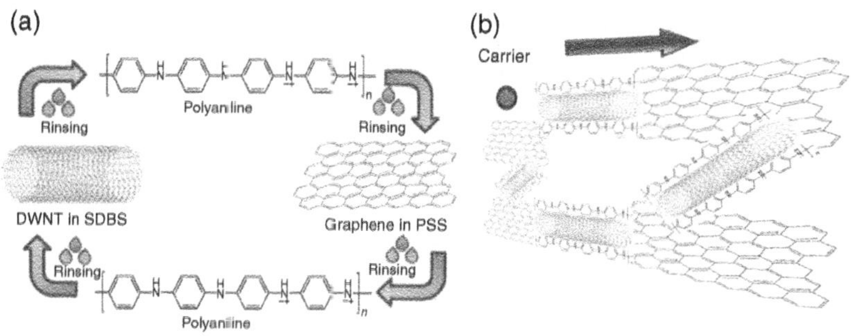

Figure 6.10 Schematic representation of (a) layer-by-layer (LBL) self-assembly procedure and (b) carrier transport in the PANI/graphene/PANi/DWNT nanocomposite multilayers. *Source:* Cho et al. [71]. © John Wiley & Sons.

Table 6.3 Summary of thermoelectric properties of TE composites using PPy or other conjugated polymers as the matrix.

Matrix	Filler	Synthesis method	Loading (wt%)	S (μVK^{-1})	σ (Scm^{-1})	PF (μWm^{-1} K^{-2})	κ (Wm^{-1}·K^{-1})	ZT	Reference
PPy	rGO	In situ chemical polymerization	50%	33.8	75.1	8.57	/	/	[33]
PPy	GNPs	In situ chemical polymerization	40%	31.74	101.68	10.24	1.09	0.0028	[32]
PPy	rGO	In situ chemical polymerization	66.67%	27	41.3	3.01	/	/	[30]
PPy	SWNT	Solution mixing	60%	26.9	300	21.7	/	/	[34]
PPy	SWNT	In situ chemical polymerization	60%	22.2	399	19.7	/	/	[76]
P3HT	MWNT	Solution mixing	5%	29.1	579	49.0	/	/	[40]
P3HT	SWNT	In situ chemical polymerization	60%	31.1	2760	267	/	/	[77]
PSB	graphite	Physical and solution mixing	80%	21	231	10.2	/	/	[74]
F8bpy	SWNT	Solution mixing	90%	46.3	290	62.3	/	/	[75]

Wang et al. [32] prepared GNP/PPy flexible materials by in-situ polymerization using $FeCl_3{\cdot}6H_2O$ as oxidant in aqueous ethanol solution (water-ethanol=1 : 1,v/v) at 0–5 °C. The GNP/PPy composite had a conductivity of 1–100 S cm^{-1}), while the conductivity increased with the increase of GNPs content. Although the GNPs exhibit high thermal conductivity, the thermal conductivity of the composite increases only slightly from 0.84 to 1.09 W $m^{-1}{\cdot}K^{-1}$ due to the entanglement with the poor thermal conductor PPy. In the end, they obtained an optimal ZT value of 2.80×10^{-3} at a GNPs to pyrrole monomer mass ratio of 1 : 0.4, which is 210 times higher than that of neat PPy.

Liang et al. [34] explored the effect of surfactant on TE performance using PPy. They first prepared PPy nanowires in cetyltrimethylammonium bromide (CTAB), which is an ammonium surfactant. The CTAB dispersed PPy is then mixed with sodium dodecyl benzene sulfonate (SDBS) or CTAB dispersed SWNTs, respectively. Thus, they obtained C/S-SWNT/PPy and S/S-SWNT/PPy films. Surfactants affect the morphology and TE properties of SWNT and PPy composites. The S/S-SWNT/PPy obtained by using SDBS surfactant showed a more dispersive morphology in SEM and TEM images, and also exhibited better thermoelectric performance than C/S-SWNT/PPy. This was attributed to reduced SWNT aggregation and increased SWNT/PPy π–π interaction sites. The composite film of S/S-SWNT/PPy achieved a PF value of 21.7 ± 0.8 μW $m^{-1}{\cdot}K^{-2}$ under an SWNT loading of 60 wt%. Similarly, Zhang et al. [33] also used CTAB as a dispersing agent to disperse rGO, and used p-toluenesulfonic acid as a dopant to polymerize PANI to obtain rGO/PPy composite film. The conductivity of the composite membrane followed the percolation law. When the rGO content was less than 25 wt%, the conductivity increases very slowly. After reaching the percolation threshold, they obtained a PF of up to 8.56 ± 0.76 μW $m^{-1}{\cdot}K^{-2}$ at 50 wt% rGO.

Considering the impact of polymerization, Aghelinejad et al. [36] investigated the effect of oxidant–monomer ratio, filler–monomer ratio, and types of polymerization additive on the thermoelectric properties of MWNT/PPy composite films. They found that the optimum molar ratio of oxidant to Py monomer was 2.5, which allowed $FeCl_3$ as an oxidant to completely polymerize PPy. Smaller oxidant-to-monomer ratios may result in fewer PPy and more porosity, which is not conducive to efficient transport path formation. At the same time, they tried to add sodium dodecyl sulfate (SDS), methyl orange (MO) and methanol during the polymerization, and found that methanol helps to form a structure with smaller porosity. Finally, they obtained the optimal PF of ~0.23 μW $m^{-1}{\cdot}K^{-2}$

with methanol addition during polymerization when the mass ratio of MWNT to Py monomer is 1.

6.2.4 Other P-Type Conducting Polymers

In addition to the abovementioned widely studied conducting polymers, there are a few other conducting polymers also used in thermoelectric materials. Poly-Schiff base is a class of organic materials having a conjugated structure and also chelating with transition metal ions. Li et al. [74] synthesized a poly-Schiff base with a donor–acceptor structure and compared with poly-Schiff base without donor–acceptor structure. Poly-Schiff base with a donor–acceptor structure has the thiophene unit and the naphthyl unit separated by a C=N bond. The naphthyl unit plays the role of a donor because of rich electronics, in contrast thiophene unit is the receptor due to electron deficiency. The composite material was obtained by mechanically mixing PSB with SWNT or Graphite in absolute ethanol. Due to the high electrical conductivity and the improved Seebeck coefficient, the PF with the donor–acceptor PSB is 10.2 $\mu W\,m^{-1}{\cdot}K^{-2}$ at a content of 80 wt% graphite, while the PF of the PSB of the common structure is 4.5 $\mu W\,m^{-1}{\cdot}K^{-2}$.

P3HT is a semiconducting polymer that has been widely used in organic photovoltaic cells since it was first synthesized in the 1980s. Qu et al. [40] used P3HT to explore the effect of the diameter of CNTs on the thermoelectric properties of composites. They prepared the SWNTs with diameters of <8, 8–15, 20–30, 30–50 and >50 nm. The two sizes of CNTs with diameters less than 15 nm exhibited better thermoelectric properties in CNT/P3HT composites. Composites with CNTs less than 8 nm in diameter had a maximum PF of 49 $\mu W\,m^{-1}{\cdot}K^{-2}$, which was attributed to complete connection of P3HT and CNT due to the good dispersion of CNTs in P3HT matrix.

Polyfluorene and its derivatives are another important class of conjugated polymeric materials. Their composites with carbon nanomaterials also demonstrate two-phase interfacial effect and π–π interactions. For example, Pan et al. [75] synthesized a bipyridyl-containing polyfluorene derivative (F8bpy) by Suzuki polymerization, and mixed the polyfluorene derivative with SWNT in chlorobenzene solution to form a composite film on the glass substrate. Figure 6.11 shows the synthesis of F8bpy and the preparation of composites with SWNT. The conductivity of the composite increases monotonously with the increase of SWNT, and the Seebeck coefficient increases with the increase of SWNT and reaches saturation at 50 wt% loading. Since F8bpy is uniformly wound on the SWNT, the carrier mobility is increased, which synergistically

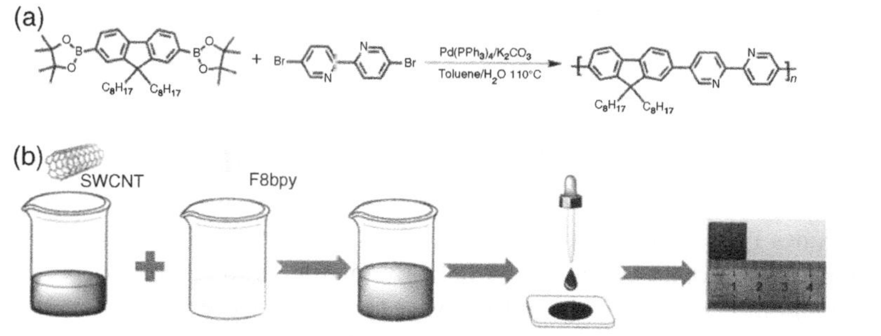

Figure 6.11 Synthesis of F8bpy (a), and schematic illustration showing the preparation process of the F8bpy or F8bpy/metal complex/SWCNT nanocomposites (b). *Source:* Pan et al. [75]. © MDPI.

increases the conductivity and the Seebeck coefficient, and achieves an optimum PF value of 62.3 $\mu W m^{-1} \cdot K^{-2}$ at 90 wt% of the SWNT load.

6.2.5 N-Type TE Composites

The development of thermoelectric materials shows that the research progress of p-type semiconductors is much faster than n-type semiconductors, but an efficient thermoelectric device needs both p-type and n-type semiconductors, so high performance n-type TE materials are indispensable in the TE research field.

CNTs are mostly p-type semiconductors because of their surface defects containing –OH or –COOH groups. But there are ways to turn them into n-type semiconductors. N-type CNTs can be achieved typically by doping with alkaline metals, nitrogen, potassium, or iodine [78, 79]. But the doping level is low and unstable in air using the chemical route. Alternatively, functionalization or wrapping the CNTs with electron donating organic compounds can achieve stable n-type CNTs in air. For example, polyethylenimine (PEI) contains many amine groups with lone pair electrons. The excess electrons on PEI can be transferred to CNTs to achieve an n-doping effect if the two materials are in close proximity. Adopting such an approach, Bark et al. dip-coated buckypapers made from SWNTs in PEI solution (SWNT-PEI) [80]. The SWNT-PEI composite demonstrated a conductivity of 110 $S cm^{-1}$ at 293 K and 180 $S cm^{-1}$ at 373 K. The Seebeck coefficient of SWNT-PEI was found to be $-50\,\mu V K^{-1}$ at 298 K, and almost unchanged at 373 K. So the PF value increased from ~30 to ~50 $\mu W m^{-1} K^{-2}$ when temperature changed from 298 to 373 K.

PEI can have different molecular structures that lead to different bulk properties. For example, linear PEI is semi-crystalline while branched PEI is amorphous. Tzounis et al. investigated the effect of PEI crystallinity on the thermoelectric properties of SWNT-PEI composites [81]. Both composites showed no weight loss up to 425 °C and excellent thermomechanical behavior. In terms of TE properties, linear PEI delivered better electrical conductivity and Seebeck coefficient than branched PEI. The highest PF values was 0.037 $\mu W m^{-1} \cdot K^{-2}$ for the linear PEI loaded with 10 vol% SWNTs.

Besides PEI, Oshima et al. explored a method to improve the stability of n-type CNTs doped by hydrazine [82]. By mixing hydrazine-doped CNTs with common polymers like poly(vinyl chloride) (PVC) or poly(vinylidene difluoride) (PVDF), the CNT/Hydrazine/Polymer (PVC or PVDF) hybrid film exhibited a S value of −52 and −48 $\mu V K^{-1}$, respectively. The negative Seebeck coefficient can be kept even after 28 days in air. In contrast, the hydrazine-treated CNTs without polymer protection changed the sign of Seebeck coefficient after 150 min due to oxidization by oxygen in air.

$NaBH_4$ as a typical reducing agent can also change CNTs from p-type to n-type. The mechanism is similar to that of hydrazine. In an article by Yu et al., PEI and NaBH4 were used together to achieve a high Seebeck coefficient of −80 $\mu V K^{-1}$ [83]. N-type semiconductor with PEI as the dopant is obtained by vacuum filtration. The n-type characteristics were maintained when the samples were left in an oxygen deficient environment or laminated with polyester films. Furthermore, the authors showed that $NABH_4$ treated CNTs exhibited a much higher electrical conductivity, compared with PEI or PEI/$NABH_4$. They also fabricated TEG with both p- and n-type CNT composites, the device could output ~6 mV thermoelectric voltage and generate ~25 nW power at a temperature gradient of 22 °C.

6.3 Non-Conducting Polymers

Although the composites of conducting polymer and CNTs have excellent thermoelectric properties, environmental stability, especially at elevated temperatures and/or high moisture, has become an important issue for conducting polymer based TE composites. Therefore, research of non-conducting polymers and carbon nanomaterials also attracted much attention. The most studied insulating polymers include polyvinylidene fluoride (PVDF) [84], polymethyl methacrylate (PMMA), polyvinyl acetate (PVA), polystyrene (PS), and polycarbonate (PC) [81, 85, 86]. Due to the change of polymer matrix, the mechanism of how non-conducting polymers and CNTs form conductive networks is also different. The TE composites using non-conducting polymers as matrix and their TE properties are summarized in Table 6.4.

Table 6.4 Summary of the TE properties of non-conducting polymer composites with carbon nanomaterials.

Polymer	Filler	Synthesismethod	Loading	S (μVK^{-1})	σ (Scm^{-1})	PF (μWm^{-1}K^{-2})	κ (Wm^{-1}K^{-1})	ZT	Reference
PEI	SWNT	Solution mixing and vacuum filtration	10vol%	43	200	0.037	/	/	[81]
PC	MWNT-COOH/CBT	Melt-mixing	2.5wt%	13.55	5	7.6×10^{-6}	0.3	7.94×10^{-9}	[86]
PDMS	MWCNTs	Solution mixing and curing	10wt%	10.4	/	/	/	7.7×10^{-7}	[87]
PVP/PEI	Graphene/DWCNT	Layer-by-layer	47wt%/ 53wt%	−80	300	190	/	/	[88]
PVDF	MWCNT	Solvent vapor-induced phase separation	15wt%	324.4	/	1.679	/	0.0033	[89]
PVAc	CNTs	Segregated network	20wt%	80	48	30.7	0.34	0.006	[37]
PVAc	DWCNT:TCPP	Segregated network	12wt%	78	71.09	42.8	0.17	/	[90]
Cellulose acetate	SWCNT	Bar-coating	90wt%	45.5	873	1.41	/	/	[91]
PS	CNT	Simple mixing	75wt%	57	2.1	0.15	/	/	[92]
PEI/$NaBH_4$	SWNT	Simple mixing	/	−80	20	12.8	/	/	[83]
Nafion	FWNT	Simple mixing	30wt%	27	12.5	1	/	/	[93]

6.3.1 Wrap

In the composites of CNTs and insulating polymers, because of the high conductivity of CNTs and the low thermal conductivity of polymers, CNTs usually form a conductive network that embedded in the polymer matrix [94]. The wrapping of CNTs by insulating polymer would change the surface property from hydrophobic to hydrophilic and can also reduce the high thermal conductivity of the CNTs network.

Choi et al. [93] compared the conductivity and Seebeck coefficient of three different kinds of CNTs, i.e. SWNTs, MWNTs, and FWNTs, when they were compounded with Nafion. Nafion is a water-soluble polymer with hydrophilic groups, so it can disperse CNTs in water without additional surfactants. At the same time, it can also be used as a polymer matrix to reduce the thermal conductivity of the whole composite. Among the three types of CNTs, the electrical performance of FWNT and MWNT was obviously superior to SWNT, but the Seebeck coefficient has no direct relationship with the type of CNTs. The highest conductivity was achieved by MWNT/Nafion ($13\,S\,cm^{-1}$); while the Seebeck coefficient of the MWNT/Nafion was $82\,\mu V\,K^{-1}$.

Suemori et al. [95] added polystyrene (PS) to SWNT, where PS was used as a binder. With the increase of PS contents, Seebeck coefficient increases and conductivity decreases. PF value reached the highest of $85\,\mu W\,m^{-1}\cdot K^{-2}$ when PS content was 20 wt%, which was 1.7 times of that of pure CNT film. The conductivity of the composite follows the percolation model. With the increase of PS content the tunneling barriers between CNT contacts increase, resulting in energy filtering effect, which enables high-energy carriers to hop and transfer, thus improving the Seebeck coefficient. The authors [92] also used printing technology to fabricate flexible and lightweight thermoelectric devices on PET substrates using CNTs/PS composites. The Seebeck coefficient was $57\,\mu V\,K^{-1}$ and electrical conductivity was $2.1\,S\,cm^{-1}$, leading to a PF of $0.15\,\mu W\,m^{-1}\cdot K^{-2}$.

Adding oligomers as dopants is also a way to improve thermoelectric properties. Tzounis et al. [86] proposed adding of cyclic butylene terephthalate (CBT) into MWNT/PC composite. The chemical structure of PC and CBT are shown in Figure 6.12a and b. CBT can react with hydroxyl groups on functionalized CNTs by transesterification, which is similar to p-type doping (Figure 6.12c). The authors claimed that such a chemical reaction could increase the density of states of MWNT, thereby increasing the Seebeck coefficient and conductivity. As the result, the Seebeck coefficient improved from $9.12 \pm 1.74\,\mu V\,K^{-1}$ for the MWNT/PC to $12.55 \pm 0.44\,\mu V\,K^{-1}$ for the MWCNT/PC/CBT, and the electrical conductivity increased by five

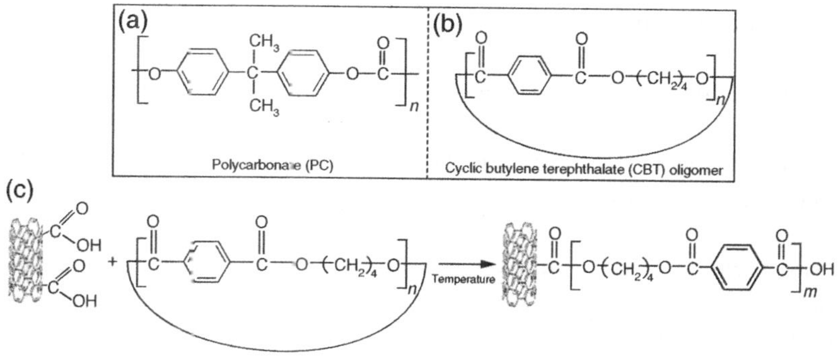

Figure 6.12 Chemical structure of (a) polycarbonate and (b) CBT oligomer (n represents the number of repeating units ranging between 2 and 7). (c) Possible chemical reaction between MWNT-COOH and CBT oligomer. *Source:* Tzounis et al. [86]. © Elsevier.

orders of magnitude, from 2.9×10^{-9} S cm^{-1} to 1.0×10^{-4} S cm^{-1}, after addition of CBT.

In CNT/non-conducting polymer systems, besides changing the non-conducting polymer or adding other materials, functionalization of CNTs can also be used to modify the composites [96]. Liebscher et al. [97] used functionalized MWNTs with -COOH and -OH groups to compound with PC matrix to form complexes. This is because the –COOH and –OH functional groups increased the carrier density in CNTs, which in turn could influence the Seebeck coefficient of the composite. When the mass fraction of CNT was 2.5%, the Seebeck coefficient of MWNT-COOH/PC reached $11.34\pm.24$ μV K^{-1}, and the pristine MWNT/PC was 7.46 ± 0.93 μV K^{-1}.

In addition to making thin films, Choi et al. [87] also made thermoelectric elastomers using poly(dimethylsiloxane) (PDMS) and CNTs. With the increase of the content of CNT, the thermoelectric properties of elastomers gradually increased. When the content of CNT increased from 6.25 to 10 wt%, the conductivity increased by 250 times. Also, from 8.8 to 10.0 wt%, the ZT values sharply increased about 20 times, and the final Seebeck coefficient and ZT value were 10.4 μV K^{-1}, 7.7×10^{-7}, respectively at 100 °C.

6.3.2 Layer-by-Layer Deposition

The application of LBL technology in thermoelectric materials can greatly increase the thermoelectric properties. Multilayer coating of PEDOT:PSS [98] and PANI/CSA [99] are good examples. The same LBL technology are also widely explored in non-conducting polymers.

Cho et al. [88] proposed to use PEI to disperse DWNT, PVP to disperse graphene, and LBL technology to assemble them into interconnected 3D nanostructures. The DWNTs bundles are interconnected and serve as bridges to graphene, so that the DWNTs and graphene are completely connected together to form a 3D conducting grid. Such a structure decouples the conductivity and the Seebeck coefficient. As the number of layers increases, more and more intensive conductive paths are formed, which greatly improves the conductivity. However, the Seebeck coefficient does not decrease significantly. This is because the energy filtering effect produced by the junction makes the average energy of carriers increase and the concentration of carriers decrease slightly, and overall the Seebeck coefficient does not change much. Moreover, graphene can prevent oxygen from penetrating into the composite, thus maintaining long-term stability in the air. Finally, the conductivity of 80 layers is $300\,S\,cm^{-1}$, the Seebeck coefficient is $-80\,\mu V\,K^{-1}$, and the PF value is $190\,\mu W\,m^{-1}{\cdot}K^{-2}$ at room temperature. Similarly, the author used PEDOT:PSS to disperse DWCNT and PANI to disperse graphene, and also used LBL technology to self-assemble films [100]. The films displayed a conductivity of $1.9{\times}x10^5$ $S\,cm^{-1}$, the Seebeck coefficient of $120\ \mu V\,K^{-1}$, and PF value of $2710\,\mu W\,m^{-1}{\cdot}K^{-2}$, which was the best TE performance at that time.

In addition to aqueous polymer, there are polymers that are soluble in organic solvents. PVDF is a popular matrix polymer but insoluble in water [101, 102]. Aghelinejad et al. [103] used LBL self-assembly technique to deposit GNPs and MWNT on the macroporous template of PVDF to form a 3D network structure. In the absence of a significant impact on Seebeck coefficient values, the conductivity is increased by nearly three orders of magnitude. Among them, when MWNTs/GNPs = 0.1, the conductivity is the highest, because MWNTs play a connecting role in GNPs and change the original 2D structure into 3D structure. Further, increasing the ratio is equal to 1, MWNTs tend to agglomerate on the surface of GNPs and play an adverse role. The thermal conductivity does not increase with the addition of conductive fillers, because there will be phonon scattering on the filler-filler junction, which reduces the thermal conductivity. On the Seebeck coefficient, because of the energy filtering effect, the Seebeck coefficient of the composite is higher than that of the pure two carbon materials. Combining the abovementioned merits, the maximum ZT value about $1{\times}10^{-3}$ was found in PVDF containing low concentration of carbon nanofillers.

Du et al. also constructed the 3D-network structure consisting of DWNT and PVDF by employing solvent vapor-induced phase separation

(SVIPS) technique [89]. By using the dimethylformamide/acetone mixture, porous networks could be generated in the MWNT/PVDF composites upon the rapid volatilization of acetone. When the precent of MWNT was 15 wt%, the highest Seebeck coefficient was 324.45 $\mu V K^{-1}$ and the highest PF and ZT value reached 1.679 $\mu W m^{-1} \cdot K^{-2}$ and 3.3×10^{-3}.

Jang et al. [104] explored the difference between CNT and GNP as conductive materials after film formation using LBL technology. GNP-based LBL films demonstrated a conductivity to be 10 $S cm^{-1}$, while the SWNT samples were increased by two orders of magnitude. Their Seebeck coefficients are similar, SWNT LBL film was −14 $\mu V K^{-1}$ while that of GNP was 15 $\mu V K^{-1}$. When the number of layer was up to 10, the SWNT LBL films possess the highest PF of 28 $\mu W m^{-1} \cdot K^{-2}$. As a one-dimensional material, CNT can form a three-dimensional structure after multi-layer deposition. In contrast, GNP is a two-dimensional material. It is difficult to form a three-dimensional structure after deposition. This explains the different electrical properties between the two carbon nanomaterials.

6.3.3 Segregated Network

In addition to the polymer encapsulated outside the CNT, CNT can also fill the gap between polymer emulsion molecules to form conductive network. This method usually adopts PVA emulsion as the polymer.

Yu et al. [37] used PVA emulsion and CNTs stabilized with Gum Arabic (GA) in aqueous solution to formulate the structure of segregated network. CNTs was connected with van der Waals force to form an orderly conducting circuit. With a CNT concentration of 20 wt%, these composites exhibit an electrical conductivity of 48 $S cm^{-1}$, thermal conductivity of 0.34 $W m^{-1} \cdot K^{-1}$ and a ZT value greater than 0.006 at room temperature.

Moriarty et al. [90] compares two CNT stabilizers, e.g. sodium deoxycholate (DOC) and meso-tetra(4-carboxyphenyl) porphine (TCPP). The mechanism of stabilizer is that the amino groups are adsorbed on the surface of CNT by electrostatic action or bond to change its hydrophilicity, thus dispersing CNTs in aqueous solution (Figure 6.13). The Seebeck coefficient is independent of the concentration of CNTs, but when DOC is replaced by TCPP, it increases three times because of the increase of effective mass and charge carriers. The Seebeck coefficient reached about 8 $\mu V K^{-1}$ for MWNT:DOC/PVA and 78 $\mu V K^{-1}$ for DWNT:TCPP/PVA. The electrical conductivity was 14.71 $S cm^{-1}$ for WMNT:DOC/PVA and 71.08 $S cm^{-1}$ for DWNT:TCPP/PVA.

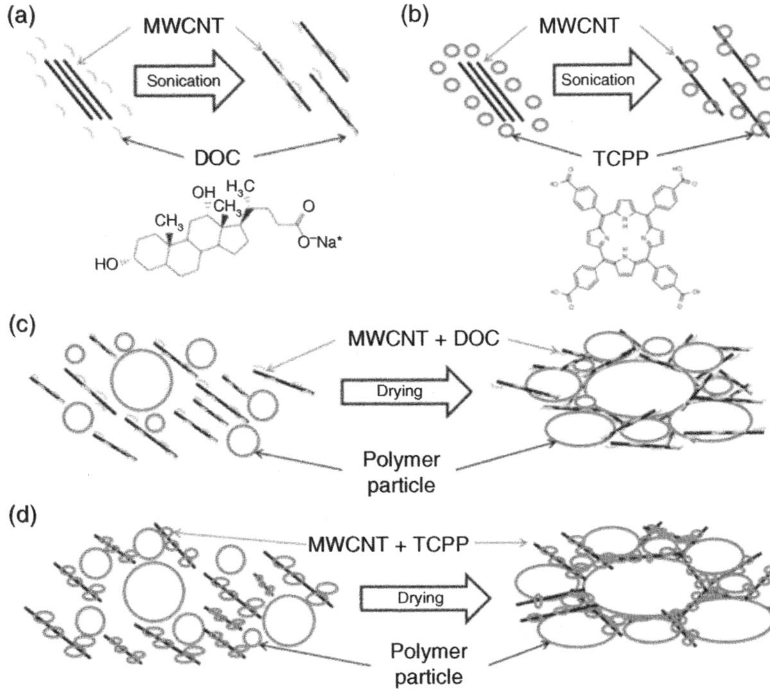

Figure 6.13 The dispersed MWCNT, when mixed with polymer emulsion particles, forms a three-dimensional network in the interstitial positions between the polymer particles (a) and (b) show representations of the dispersed MWCNT in the two stabilizing agents, DOC and TCPP, (c) and (d) illustrate the formation of a segregated network upon drying of the water-based polymer emulsions. *Source:* Moriarty et al. [90]. © Elsevier.

6.4 Ternary Thermoelectric Material

Although the composite of polymer and carbon material has a significant impact on thermoelectric properties, existing work has shown that further addition of inorganic nanoparticles improves the thermoelectric properties. Different nanostructures brought by ternary composites further affect thermoelectric properties.

6.4.1 Non-conducting Polymer

In order to facilitate better dispersion of CNTs, Toshima et al. [105] added an n-type semi-conducting polymer complex to the CNT/PVA binary system. The n-type semi-conducting polymer complex consists of nickel metal and poly(1,1,2,2-ethenetetrathiolate), abbreviated as *n*-PETT. The nanoparticles of *n*-PETT can be well dispersed in PVC hybrid. And it plays an important role in smoothing the CNT tube-tube contact and thus results in better charge transport between CNT bundles in the films. According to SEM images, the CNT bundle size decreases from 10.9 ± 5.4 to 6.1 ± 3.6 nm with the addition of n-PETT. As a result, the conductivity increases dramatically, from 23.2 ± 5.2 to $429.3 \pm 7.1\,\mathrm{S\,cm^{-1}}$, and the Seebeck coefficient does not change much, from 31.9 ± 2.8 to $30.1 \pm 0.6\,\mathrm{\mu V\,K^{-1}}$.

Sepahvand et al. introduced a new method for preparing nanocomposites containing metal nanoparticles, CNTs and polymer [106]. In this method, the structure of CNTs was opened by acid treatment. Silver nanoparticles were introduced into the CNTs via wet chemistry method. Then MWNT containing metal nanoparticles were used as macro-initiator for ring opening polymerization of caprolactone and MWNT/Ag nanoparticles/polymer ternary nanocomposites were obtained. Unfortunately the authors did not measure the TE properties of such a composite materials.

6.4.2 Conducting Polymer

PEDOT/rGO/SWNT composites with good thermoelectric properties were successfully prepared by Li et al. [38] through in situ polymerization and physical mixing. Figure 6.14 illustrates the fabrication process of the ternary composite. The SEM image shows that the PEDOT-coated rGO nanoplates and SWCNT were uniformly dispersed to form a conductive network. Compared with pure PEDOT and PEDOT/rGO binary composites, the ternary composites have the highest conductivity, which increases with the increase of SWNT loading. Although the Seebeck coefficient of the ternary composite was higher than that of pure PEDOT, its value was similar to that

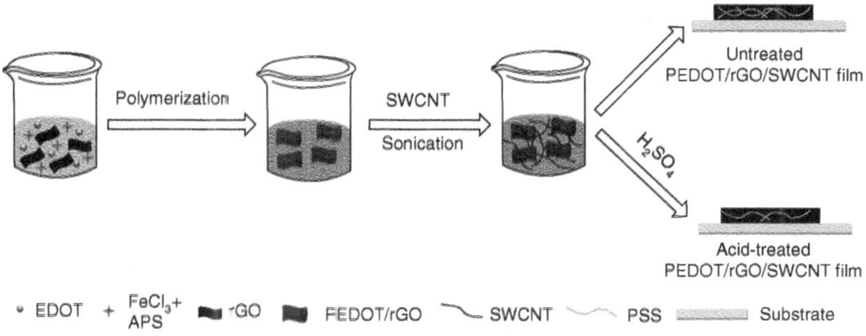

Figure 6.14 Schematic illustration of the preparation procedure of the PEDOT/rGO/SWNT ternary composites with enhanced thermoelectric performance. *Source:* Li et al. [38]. © Elsevier.

of the binary composite of rGO/PEDOT. Furthermore, acid treatment can significantly improve the electrical conductivity of pure PEDOT and all composites, but will lead to a decrease in the Seebeck coefficient. In general, the maximum PF of PEDOT/rGO/SWNT was $9.10 \pm 0.62\,\mu W\,m^{-1}\cdot K^{-2}$ after acid treatment at 10 wt% SWNT.

Meng et al. [107] prepared SWNT/PEDOT:PSS-coated Te nanorods (PC-Te) composite membranes by a simple vacuum-assisted filtration method (Figure 6.15). The addition of SWNT improves the conductivity obviously, and reduces the Seebeck coefficient of the composite film, which leads to the reduction of the PF. After H_2SO_4 treatment, the conductivity of the composite film was further improved, and the Seebeck coefficient decreased slightly. Thus, the sample containing 70 wt% SWNT demonstrates an optimized PF of $104\,\mu W\,m^{-1}\cdot K^{-2}$ at 300 K, corresponding to a conductivity of $332\,S\,cm^{-1}$ and a Seebeck coefficient of $56\,\mu V\,K^{-1}$, respectively.

In some cases, metals such as Au, Ag, Te, etc. are added to the thermoelectric composite due to intrinsically good electrical conductivity. Cao et al. [108] used graphene oxide (GO) as surfactant for CNT and decorated with Ag to prepare Ag-GO/CNT/PANI composites by chemical oxidization. Compared with pure PANI, this composites had higher Seebeck coefficient because of the energy filtering effect between CNT and PANI matrix, leading to the enhancement of the density of the states at the Fermi level. The lattice thermal conductivity was kept at a low level by the scattering of

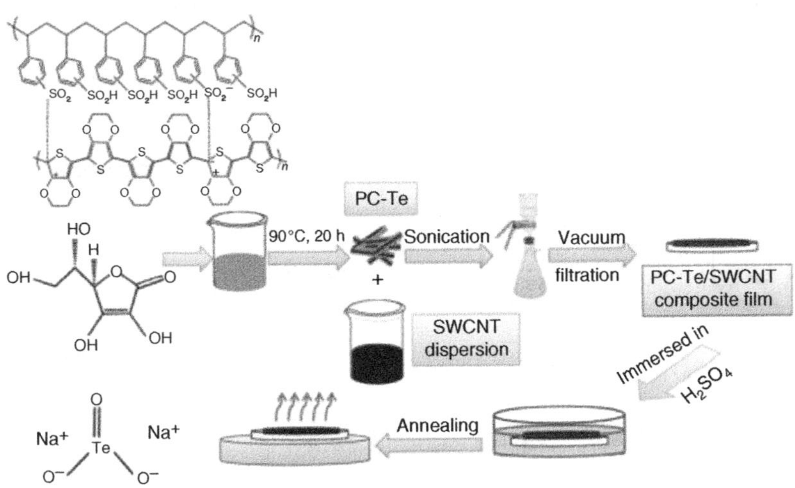

Figure 6.15 Schematic illustration of the fabrication and post-treatment of the flexible SWNT/PC-Te composites films. *Source:* Meng et al. [107]. © Elsevier.

phonons at the interface. The functionalization of GO and CNT introduced defects to the carbon nanomaterials, and reduced their intrinsic conductivity. These defects also increased chances for carrier scattering when PANI was polymerized on these materials, thus further reducing the conductivity. To remedy the problem, the addition of highly-conductive silver can improve the conductivity. The PF of Ag-GO/CNT/PANI composite reached up to $0.32\,\mu W\,m^{-1}{\cdot}K^{-2}$ at 120 °C, which was 71 and 91% higher than PANI and CNT/PANI, respectively.

Similarly, An et al. [109] prepared foldable TE materials by embedding PANI or P3HT into Au-doped CNT webs. They first formed a free-standing CNT web in a heating furnace, and then immersed the CNT network in $AuCl_3$ solution and dried. Au^{3+} was reduced to metallic Au by withdrawing electrons from the CNT, because it had a lower reduction potential, resulting in the formation of well-scattered Au particles on the CNT webs. Finally, ternary TE composite could be obtained by vacuum filtration of PANI or P3HT through the CNT web inlaid with Au particles. The good conductivity of Au greatly improves the electrical conductivity of the composite film, which can reach $1106\,S\,cm^{-1}$ at room temperature. In addition, the introduction of Au creates new interfaces, which can effectively scatter the phonons, while the thermal conductivity is only slightly increased. The energy filtering effect caused by PANI-coated Au-modified CNT web results in a high Seebeck coefficient value of $150.86\,\mu V\,K^{-1}$. The ZT value of 0.023 is the highest value they obtained.

In 2016, Wang et al. introduced Te nanorods in PANI/SWNT materials to prepare ternary composite films [5]. PANI is used as a shell to wrap SWNT and Te nanorods as nuclei. The interfaces of PANI/SWNT (0.47 eV) and PANI/Te (0.32 eV) have different energy barriers, which effectively scatter carriers with low energy causing enhanced Seebeck coefficient. The maximum PF of the film is up to $101\,\mu W\,m^{-1}{\cdot}K^{-2}$.

In addition to the introduction of highly conductive metal particles among carbon fillers and polymers, the formation of complexes by the incorporation of bipyridine units to chelate metal ions on the polymer backbone is also a method to form ternary composites. Pan et al. [75] chelated transition metal ions by attaching a bipyridyl unit to polyfluorene derivative (F8bpy), and formed a ternary composite with SWNT. They tried Mn, Fe, Co, Ni, Cu, and Zn metal ions and the thermoelectric properties of chelated different transition metals are shown in Figure 6.16. A ternary composite with a transition metal has a higher conductivity than a binary composite of SWNT/F8bpy because the transition metal is considered to be a chemical dopant of the polymer backbone. Chelate-introduced transition metal increased the carrier concentration, caused a decrease in the Seebeck

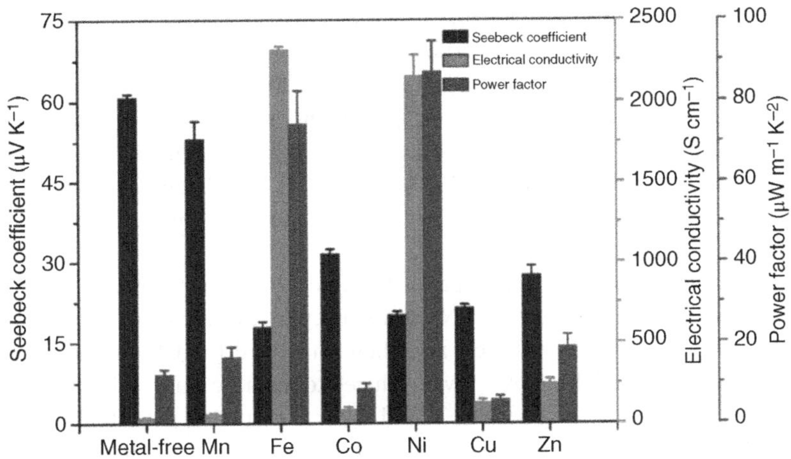

Figure 6.16 Thermoelectric properties of F8bpy/metal complex/SWNT composite films with different transition metal ions (SWNT 50 wt%). *Source:* Pan et al. [75]. © MDPI.

coefficient. However, thermoelectric performance is a comprehensive assessment of Seebeck coefficient, conductivity and thermal conductivity. Ni/SWNT/F8bpy ternary composite exhibits a conductivity of $2153.4\,S\,cm^{-1}$ and a PF of $87.3\,\mu W\,m^{-1}{\cdot}K^{-2}$ at room temperature under a load of 50 wt% SWNT.

6.5 Summary and Outlook

This review focuses on recent works involving the thermoelectric composite materials that contain carbon nanomaterials as the fillers in polymer matrix. The polymers are classified into conducting polymers and non-conductive polymers. Typical carbon-based filler materials include CNTs, graphite, graphene nanoplatelet, reduced graphene oxide, etc. The addition of carbon materials into polymer matrix can introduce multiple effect, e.g. it creates an electrically conducting network; it introduces new interfaces for phonon scattering or scattering of charge carriers with low energy. Manipulating the charge carrier concentration, mobility, molecular orientation in conducting polymer matrix are also a feasible way to fine tune the thermoelectric properties. Different fabrication and treatment techniques such as doping/de-doping, in-situ polymerization, layer-by-layer deposition, vacuum filtration, etc. can greatly influence the microstructure and the TE properties of the composite films. Despite all these effort, the dimensionless figure of merit (ZT) value of all the reviewed composites is still below 1, implying that there is still much room for achieving higher ZT value. A few directions might be considered:

1) Unlike inorganic TE materials, the fundamental understanding on charge carrier and phonon transport in the polymeric composites is still lacking. Theoretical calculation of the molecular structure of thermoelectric materials and the transport theory of composite materials are urgently needed for guiding materials design.
2) The three main TE parameters such as Seebeck coefficient, electrical conductivity and thermal conductivity are often interrelated. Even though a few examples demonstrated decoupling of the Seebeck coefficient and electrical conductivity, it is still rare to observe significant enhancement of these two parameters. Finding an effective way to greatly improve all the three thermoelectric parameters will be of great significance.
3) Recently some studies have shown ionic conductors possess a huge Seebeck coefficient in comparison to electronic conductors. But the electrical properties of ionic conductors are usually poor. If one can improve the electrical property of ionic conductors, or couple the ionic and electronic transport in hybrid materials, a drastic enhancement of TE property can be expected.

References

1 Kroon, R., Mengistie, D.A. Kiefer, D. et al. (2016). *Thermoelectric plastics: from design to synthesis, processing and structure-property relationships. Chemical Society Reviews* 45 (22): 6147–6164.

2 Dubey, N. and Leclerc, M. (2011). *Conducting polymers: efficient thermoelectric materials. Journal of Polymer Science Part B-Polymer Physics* 49 (7): 467–475.

3 Meer, S., Kausar, A., and Iqbal, T. (2016). *Trends in conducting polymer and hybrids of conducting polymer/carbon nanotube: a review. Polymer-Plastics Technology and Engineering* 55 (13): 1416–1440.

4 Zhang, Y., Heo, Y.-J., Park, M. et al. (2019). *Recent advances in organic thermoelectric materials: principle mechanisms and emerging carbon-based green energy materials. Polymers* 11 (1): 167.

5 Wang, L.M., Yao, Q., Shi, W. et al. (2017). *Engineering carrier scattering at the interfaces in polyaniline based nanocomposites for high thermoelectric performances. Materials Chemistry Frontiers* 1 (4): 741–748.

6 Ju, H., Park, D., and Kim, J. (2018). *Solution-processable flexible thermoelectric composite films based on conductive polymer/ SnSe0.8S0.2 nanosheets/carbon nanotubes for wearable electronic applications. Journal of Materials Chemistry A* 6 (14): 5627–5634.

7 Chakraborty, P., Ma, T.F., Zahiri, A.H. et al. (2018). *Carbon-based materials for thermoelectrics. Advances in Condensed Matter Physics.*

8 Wang, Q., Yao, Q., Chang, J. et al. (2012). *Enhanced thermoelectric properties of CNT/PANI composite nanofibers by highly orienting the arrangement of polymer chains. Journal of Materials Chemistry* 22 (34): 17612–17618.

9 Yao, Q., Wang, Q., Wang, L.M. et al. (2014). *Abnormally enhanced thermoelectric transport properties of SWNT/PANI hybrid films by the strengthened PANI molecular ordering. Energy & Environmental Science* 7 (11): 3801–3807.

10 Chen, J.K., Wang, L.M., Ren, D.D. et al. (2018). *Revealing the anisotropy in thermoelectric transport performances in CNT/PANI composites. Synthetic Metals* 239: 13–21.

11 Zhao, Y., Tang, G.S., Yu, Z.Z. et al. (2012). *The effect of graphite oxide on the thermoelectric properties of polyaniline. Carbon* 50 (8): 3064–3073.

12 Pan, C., Zhang, L., Pan, Z. et al. (2015). *A simple strategy to fabricate polyaniline/expanded graphite composites with improved power factor. Materials Chemistry and Physics* 167: 315–319.

13 Xiang, J.L. and Drzal, L.T. (2012). *Templated growth of polyaniline on exfoliated graphene nanoplatelets (GNP) and its thermoelectric properties. Polymer* 53 (19): 4202–4210.

14 Abad, B., Alda, I., Diaz-Chao, P. et al. (2013). *Improved power factor of polyaniline nanocomposites with exfoliated graphene nanoplatelets (GNPs). Journal of Materials Chemistry A* 1 (35): 10450–10457.

15 Xiang, J.L. and Drzal, L.T. (2014). *Improving thermoelectric properties of graphene/polyaniline paper by folding. Chemical Physics Letters* 593: 109–114.

16 Mitra, M., Ahamed, S.T., Ghosh, A. et al. (2019). *Polyaniline/reduced graphene oxide composite-enhanced visible-light-driven photocatalytic activity for the degradation of organic dyes. Acs Omega* 4 (1): 1623–1635.

17 Islam, R., Chan-Yu-King, R., Brun, J.F. et al. (2014). *Transport and thermoelectric properties of polyaniline/reduced graphene oxide nanocomposites. Nanotechnology* 25 (47): 475705.

18 Yao, H., Fan, Z., Cheng, H. et al. (2018). *Recent development of thermoelectric polymers and composites. Macromolecular Rapid Communications* 39 (6): 1700727.

19 Dey, A., Bajpai, O.P., Sikder, A.K. et al. (2016). *Recent advances in CNT/graphene based thermoelectric polymer nanocomposite: a proficient move towards waste energy harvesting. Renewable and Sustainable Energy Reviews* 53: 653–671.

20 Chen, G., Xu, W., and Zhu, D. (2017). *Recent advances in organic polymer thermoelectric composites. Journal of Materials Chemistry C* 5 (18): 4350–4360.

21 Li, Y., Du, Y., Dou, Y. et al. (2017). *PEDOT-based thermoelectric nanocomposites – a mini-review. Synthetic Metals* 226: 119–128.

22 Zhang, L., Du, W.Y., Nautiyal, A. et al. (2018). *Recent progress on nanostructured conducting polymers and composites: synthesis, application and future aspects. Science China-Materials* 61 (3): 303–352.

23 Chen, J., Gui, X., Wang, Z. et al. (2012). *Superlow thermal conductivity 3D carbon nanotube network for thermoelectric applications. ACS Appl Mater Interfaces* 4 (1): 81–86.

24 Wang, L.M., Yao, Q., Bi, H. et al. (2015). *PANI/graphene nanocomposite films with high thermoelectric properties by enhanced molecular ordering. Journal of Materials Chemistry A* 3 (13): 7086–7092.

25 Song, H.J., Cai, K.F., Wang, J. et al. (2016). *Influence of polymerization method on the thermoelectric properties of multi-walled carbon nanotubes/polypyrrole composites. Synthetic Metals* 211: 58–65.

26 Chen, J.K., Wang, L.M., Gui, X.C. et al. (2017). *Strong anisotropy in thermoelectric properties of CNT/PANI composites. Carbon* 114: 1–7.

27 Erden, F., Li, H., Wang, X. et al. (2018). *High-performance thermoelectric materials based on ternary TiO2/CNT/PANI composites. Phys Chem Chem Phys* 20 (14): 9411–9418.

28 Li, H., Liu, S.Q., Li, P.C. et al. (2018). *Interfacial control and carrier tuning of carbon nanotube/polyaniline composites for high thermoelectric performance. Carbon* 136: 292–298.

29 Lin, Y.H., Lee, T.C., Hsiao, Y.S. et al. (2018). *Facile synthesis of diamino-modified graphene/polyaniline semi-interpenetrating networks with practical high thermoelectric performance. ACS Appl Mater Interfaces* 10 (5): 4946–4952.

30 Han, S.B., Zhai, W.T., Chen, G.M. et al. (2014). *Morphology and thermoelectric properties of graphene nanosheets enwrapped with polypyrrole. Rsc Advances* 4 (55): 29281–29285.

31 Liu, J. and Yu, H.Q. (2014). *Thermoelectric enhancement in polyaniline composites with polypyrrole-functionalized multiwall carbon nanotubes. Journal of Electronic Materials* 43 (4): 1181–1187.

32 Wang, L.Y., Liu, F.Z., Jin, C. et al. (2014). *Preparation of polypyrrole/graphene nanosheets composites with enhanced thermoelectric properties. Rsc Advances* 4 (86): 46187–46193.

33 Zhang, Z., Chen, G.M., Wang, H.F. et al. (2015). *Enhanced thermoelectric property by the construction of a nanocomposite 3D interconnected architecture consisting of graphene nanolayers sandwiched by polypyrrole nanowires. Journal of Materials Chemistry C* 3 (8): 1649–1654.

34 Liang, L.R., Chen, G.M., and Guo, C.Y. (2016). *Enhanced thermoelectric performance by self-assembled layered morphology of polypyrrole nanowire/*

single-walled carbon nanotube composites. Composites Science and Technology 129: 130–136.

35 Xin, S.C., Yang, N., Gao, F. et al. (2018). *Free-standing and flexible polypyrrole nanotube/reduced graphene oxide hybrid film with promising thermoelectric performance. Materials Chemistry and Physics* 212: 440–445.

36 Aghelinejad, M., Zhang, Y.C., and Leung, S.N. (2019). *Processing parameters to enhance the electrical conductivity and thermoelectric power factor of polypyrrole/multi-walled carbon nanotubes nanocomposites. Synthetic Metals* 247: 59–66.

37 Yu, C., Kim, Y.S., Kim, D. et al. (2008). *Thermoelectric behavior of segregated-network polymer nanocomposites. Nano Letters* 8 (12): 4428–4432.

38 Li, X., Liang, L.R., Yang, M.Z. et al. (2016). *Poly(3,4-ethylenedioxythiophene)/graphene/carbon nanotube ternary composites with improved thermoelectric performance. Organic Electronics* 38: 200–204.

39 Sanchez, J.A.L., Capilla, R.P., and Diez-Pascual, A.M. (2018). *High-performance PEDOT:PSS/hexamethylene diisocyanate-functionalized graphene oxide nanocomposites: preparation and properties. Polymers*: 10(10).

40 Qu, S.Y., Wang, M.D., Chen, Y.L. et al. (2018). *Enhanced thermoelectric performance of CNT/P3HT composites with low CNT content. Rsc Advances* 8 (59): 33855–33863.

41 Zou, Y., Huang, D.Z., Meng, Q. et al. (2018). *Correlation between Seebeck coefficient and transport energy level in poly (3-hexylthiophene). Organic Electronics* 56: 125–128.

42 Gao, C.Y. and Chen, G.M. (2016). *A new strategy to construct thermoelectric composites of SWCNTs and poly-Schiff bases with 1,4-diazabuta-1,3-diene structures acting as bidentate-chelating units. Journal of Materials Chemistry A* 4 (29): 11299–11306.

43 Li, J.J., Lai, C.H., Xiang, X.Z. et al. (2015). *Synthesis and characterization of poly-Schiff bases with a donor-acceptor structure containing thiophene units as thermoelectric materials. Journal of Materials Chemistry C* 3 (11): 2693–2701.

44 Song, H.J., Liu, C.C., Xu, J.K. et al. (2013). *Fabrication of a layered nanostructure PEDOT:PSS/SWCNTs composite and its thermoelectric performance. Rsc Advances* 3 (44): 22065–22071.

45 See, K.C., Feser, J.P., Chen, C.E. et al. (2010). *Water-processable polymer–nanocrystal hybrids for thermoelectrics. Nano Letters* 10 (11): 4664–4667.

46 Kim, D., Kim, Y., Choi, K. et al. (2010). *Improved thermoelectric behavior of nanotube-filled polymer composites with poly(3,4-ethylenedioxythiophene) poly(styrenesulfonate). Acs Nano* 4 (1): 513–523.

47 Yu, C., Choi, K., Yin, L. et al. (2011). *Light-weight flexible carbon nanotube based organic composites with large thermoelectric power factors. Acs Nano* 5 (10): 7885–7892.

48 Lee, W., Kang, Y.H., Lee, J.Y. et al. (2016). *Improving the thermoelectric power factor of CNT/PEDOT:PSS nanocomposite films by ethylene glycol treatment. Rsc Advances* 6 (58): 53339–53344.

49 Choi, K., Kim, S.L., Yi, S.I. et al. (2018). *Promoting dual electronic and ionic transport in PEDOT by embedding carbon nanotubes for large thermoelectric responses. ACS Appl Mater Interfaces* 10 (28): 23891–23899.

50 Kim, G.H., Hwang, D.H., and Woo, S.I. (2012). *Thermoelectric properties of nanocomposite thin films prepared with poly(3,4-ethylenedioxythiophene) poly(styrenesulfonate) and graphene. Physical Chemistry Chemical Physics* 14 (10): 3530–3536.

51 Li, F.Y., Cai, K.F., Shen, S. et al. (2014). *Preparation and thermoelectric properties of reduced graphene oxide/PEDOT:PSS composite films. Synthetic Metals* 197: 58–61.

52 Yoo, D., Kim, J., and Kim, J.H. (2014). *Direct synthesis of highly conductive poly(3,4-ethylenedioxythiophene):poly(4-styrenesulfonate) (PEDOT:PSS)/ graphene composites and their applications in energy harvesting systems. Nano Research* 7 (5): 717–730.

53 Xiong, J.H., Jiang, F.X., Shi, H. et al. (2015). *Liquid exfoliated graphene as dopant for improving the thermoelectric power factor of conductive PEDOT:PSS nanofilm with hydrazine treatment. Acs Applied Materials & Interfaces* 7 (27): 14917–14925.

54 Xu, K., Chen, G., and Qiu, D. (2013). *Convenient construction of poly(3,4-ethylenedioxythiophene)–graphene pie-like structure with enhanced thermoelectric performance. Journal of Materials Chemistry A* **1** (40): 12395–12399.

55 Lee, W., Kang, Y.H., Lee, J.Y. et al. (2017). *Hot-pressing for improving performance of CNT/conjugated polymer thermoelectric films and power generators. Materials Today Communications* 10: 41–45.

56 Moriarty, G.P., Briggs, K., Stevens, B. et al. (2013). *Fully organic nanocomposites with high thermoelectric power factors by using a dual-stabilizer preparation. Energy Technology* 1 (4): 265–272.

57 Mannayil, J., Raman, S.M., Sankaran, J. et al. (2018). *Solution processable PEDOT:PSS/multiwalled carbon nanotube composite films for flexible electrode applications. physica status solidi (a)* 215 (18): 1701003.

58 Hsu, J.-H., Choi, W., Yang, G. et al. (2017). *Origin of unusual thermoelectric transport behaviors in carbon nanotube filled polymer composites after solvent/acid treatments. Organic Electronics* 45: 182–189.

59 Song, H., Qiu, Y., Wang, Y. et al. (2017). *Polymer/carbon nanotube composite materials for flexible thermoelectric power generator. Composites Science and Technology* 153: 71–83.

60 Moriarty, G.P., De, S., King, P.J. et al. (2013). *Thermoelectric behavior of organic thin film nanocomposites. Journal of Polymer Science Part B: Polymer Physics* 51 (2): 119–123.

61 Hu, X., Chen, G., and Wang, X. (2017). *An unusual coral-like morphology for composites of poly(3,4-ethylenedioxythiophene)/carbon nanotube and the enhanced thermoelectric performance. Composites Science and Technology* 144: 43–50.

62 Kymakis, E. and G. Amaratunga, *Electrical properties of single-wall carbon nanotube-polymer composite films. Journal of Applied Physics*, 2006. 99(8): p. 84302–0.

63 Das, N.C., Liu, Y., Yang, K. et al. (2010). *Single-walled carbon nanotube/ poly(methyl methacrylate) composites for electromagnetic interference shielding. Polymer Engineering & Science* 49 (8): 1627–1634.

64 Jiang, Q., Lan, X., Liu, C. et al. (2018). High-performance hybrid organic thermoelectric SWNTs/PEDOT:PSS thin-films for energy harvesting. *Materials Chemistry Frontiers* 2: 679–685. https://doi.org/10.1039/C7QM00515F.

65 Wang, L.M., Yao, Q., Bi, H. et al. (2014). *Large thermoelectric power factor in polyaniline/graphene nanocomposite films prepared by solution-assistant dispersing method. Journal of Materials Chemistry A* 2 (29): 11107–11113.

66 Zhang, K., Davis, M., Qiu, J. et al. (2012). *Thermoelectric properties of porous multi-walled carbon nanotube/polyaniline core/shell nanocomposites. Nanotechnology* 23 (38): 385701.

67 Wang, H., Yi, S.I., and Yu, C. (2016). *Engineering electrical transport at the interface of conjugated carbon structures to improve thermoelectric properties of their composites. Polymer* 97: 487–495.

68 Zhang, Q.L., Wang, W.J., Li, J.L. et al. (2013). *Preparation and thermoelectric properties of multi-walled carbon nanotube/polyaniline hybrid nanocomposites. Journal of Materials Chemistry A* 1 (39): 12109–12114.

69 Wang, H., Yi, S.I., Pu, X. et al. (2015). *Simultaneously improving electrical conductivity and thermopower of polyaniline composites by utilizing carbon nanotubes as high mobility conduits. ACS Appl Mater Interfaces* 7 (18): 9589–9597.

70 Lu, Y., Song, Y., and Wang, F.P. (2013). *Thermoelectric properties of graphene nanosheets-modified polyaniline hybrid nanocomposites by an in situ chemical polymerization. Materials Chemistry and Physics* 138 (1): 238–244.

71 Cho, C., Stevens, B., Hsu, J.H. et al. (2015). *Completely organic multilayer thin film with thermoelectric power factor rivaling inorganic tellurides. Adv Mater* 27 (19): 2996–3001.

72 Liu, J., Zhang, L.M., He, L. et al. (2003). *Synthesis and thermoelectric properties of polyaniline. Journal of Wuhan University of Technology-Materials Science Edition* 18 (3): 53–55.

73 Huang, P., Zhu, H.R., Jing, L. et al. (2011). *Graphene covalently binding aryl groups: conductivity increases rather than decreases. Acs Nano* 5 (10): 7945–7949.

74 Sun, K., Chang, J., Isikgor, F. et al. (2015). *Efficiency enhancement of planar perovskite solar cells by adding zwitterion/LiF double interlayers for electron collection. Nanoscale* 7: 896–900.

75 Pan, C.J., Wang, L.H., Zhou, W.Q. et al. (2019). *Preparation and thermoelectric properties study of Bipyridine-containing polyfluorene derivative/SWCNT composites. Polymers* 11 (2): 278.

76 Liang, L., Gao, C., Chen, G. et al. (2016). *Large-area, stretchable, super flexible and mechanically stable thermoelectric films of polymer/carbon nanotube composites. Journal of Materials Chemistry C* 4 (3): 526–532.

77 Hong, C.T., Lee, W., Kang, Y.H. et al. (2015). *Effective doping by spin-coating and enhanced thermoelectric power factors in SWCNT/P3HT hybrid films. Journal of Materials Chemistry A* 3 (23): 12314–12319.

78 Park, T., Sim, K., Lee, J. et al. (2012). *Electron transfer properties of iodine-doped single-walled carbon nanotubes using field effect transistor.* 12 (7): 5812–5815.

79 Kim, B.H., Park, T.H., Baek, S.J. et al. (2008). *N-type carbon nanotube by alkaline-earth metal Sr doping. AIP*: 096103.

80 Bark, H., Lee, W., and Lee, H. (2015). *Enhanced thermoelectric performance of CNT thin film p/n junctions doped with N-containing organic molecules. Macromolecular Research* 23 (9): 795–801.

81 Tzounis, L., Hegde, M., Liebscher, M. et al. (2018). *All-aromatic SWCNT-Polyetherimide nanocomposites for thermal energy harvesting applications. Composites Science and Technology* 156: 158–165.

82 Oshima, K., Yanagawa, Y. Asano, H. et al. (2017). *Improvement of stability of n-type super growth CNTs by hybridization with polymer for organic hybrid thermoelectrics. Synthetic Metals* 225: 81–85.

83 Yu, C.H., Murali, A., Choi, K.W. et al. (2012). *Air-stable fabric thermoelectric modules made of N- and P-type carbon nanotubes. Energy & Environmental Science* 5 (11): 9481–9486.

84 Montgomery, D.S., Hewitt, C.A., Barbalace, R. et al. (2016). *Spray doping method to create a low-profile high-density carbon nanotube thermoelectric generator. Carbon* 96: 778–781.

85 Piao, M.X., Alam, M.R., Kim, G. et al. (2012). *Effect of chemical treatment on the thermoelectric properties of single walled carbon nanotube networks. Physica Status Solidi B-Basic Solid State Physics* 249 (12): 2353–2356.

86 Tzounis, L., Gartner, T., Liebscher, M. et al. (2014). *Influence of a cyclic butylene terephthalate oligomer on the processability and thermoelectric properties of polycarbonate/MWCNT nanocomposites. Polymer* 55 (21): 5381–5388.

87 Choi, J.H., Hyun, C.M., Jo, H. et al. (2017). *Thermoelectric elastomer fabricated using carbon nanotubes and nonconducting polymer. Japanese Journal of Applied Physics* **56** (9): 095101.

88 Cho, C., Yu, C., and Grunlan, J. (2016). *Stable n-type thermoelectric multilayer thin films with high power factor from carbonaceous nanofillers. Nano Energy* 28: 426–432.

89 Du, F.P., Qiao, X., Wu, Y.G. et al. (2018). *Fabrication of porous polyvinylidene fluoride/multi-walled carbon nanotube nanocomposites and their enhanced thermoelectric performance. Polymers* 10 (7): 797.

90 Moriarty, G.P., Wheeler, J.N., Yu, C.H. et al. (2012). *Increasing the thermoelectric power factor of polymer composites using a semiconducting stabilizer for carbon nanotubes. Carbon* 50 (3): 885–895.

91 Mo, J.H., Kim, J.Y., Kang, Y.H. et al. (2018). *Carbon nanotube/cellulose acetate thermoelectric papers. Acs Sustainable Chemistry & Engineering* 6 (12): 15970–15975.

92 Suemori, K., Hoshino, S., and Kamata, T. (2013). *Flexible and lightweight thermoelectric generators composed of carbon nanotube-polystyrene composites printed on film substrate. Applied Physics Letters* 103 (15): 153902.

93 Choi, Y., Kim, Y., Park, S.G. et al. (2011). *Effect of the carbon nanotube type on the thermoelectric properties of CNT/Nafion nanocomposites. Organic Electronics* 12 (12): 2120–2125.

94 Asano, H., Sakura, N., Oshima, K. et al. (2016). *Development of ethenetetrathiolate hybrid thermoelectric materials consisting of cellulose acetate and semiconductor nanomaterials. Japanese Journal of Applied Physics* 55 (2): 02BB02.

95 Suemori, K., Watanabe, Y., and Hoshino, S. (2016). *Increase in thermoelectric power factor of carbon-nanotube films after addition of polystyrene. Organic Electronics* 28: 135–138.

96 Moniruzzaman, M. and Winey, K.I. (2006). *Polymer nanocomposites containing carbon nanotubes. Macromolecules* 39 (16): 5194–5205.

97 Liebscher, M., Gartner, T., Tzounis, L. et al. (2014). *Influence of the MWCNT surface functionalization on the thermoelectric properties of melt-mixed polycarbonate composites. Composites Science and Technology* 101: 133–138.

98 Andrei, V., Bethke, K., Madzharova, F. et al. (2017). *Size dependence of electrical conductivity and thermoelectric enhancements in spin-coated PEDOT:PSS single and multiple layers. Advanced Electronic Materials* 3 (2): 1600473.

99 Andrei, V., Bethke, K., Madzharova, F. et al. (2017). *In situ complementary doping, thermoelectric improvements, and strain-induced structure within alternating PEDOT:PSS/PANI layers. Acs Applied Materials & Interfaces* 9 (38): 33308–33316.

100 Cho, C., Wallace, K.L., Tzeng, P. et al. (2016). *Outstanding low temperature thermoelectric power factor from completely organic thin films enabled by multidimensional conjugated nanomaterials. Advanced Energy Materials* 6 (7): 1502168.

101 Kim, J.Y., Mo, J.H., Kang, Y.H. et al. (2018). *Thermoelectric fibers from well-dispersed carbon nanotube/poly(vinyliedene fluoride) pastes for fiber-based thermoelectric generators. Nanoscale* 10 (42): 19766–19773.

102 Yoon, H.D., Nam, S., Tu, N.D.K. et al. (2013). *Thermoelectric properties of graphite nanosheets/poly(vinylidene fluoride) composites. Polymer-Korea* 37 (5): 638–641.

103 Aghelinejad, M. and Leung, S.N. (2018). *Thermoelectric nanocomposite foams using non-conducting polymers with hybrid 1D and 2D nanofillers. Materials* 11 (9): 1757.

104 Jang, W., Cho, H.A., Choi, K. et al. (2018). *Manipulation of p–/n-type thermoelectric thin films through a layer-by-layer assembled carbonaceous multilayer structure. Micromachines* 9 (12): 628.

105 Toshima, N., Oshima, K., Anno, H. et al. (2015). *Novel hybrid organic thermoelectric materials. three-component hybrid films consisting of a nanoparticle polymer complex, carbon nanotubes, and vinyl polymer. Advanced Materials* 27 (13): 2246-+.

106 Sepahvand, R., Adeli, M., Astinchap, B. et al. (2008). *New nanocomposites containing metal nanoparticles, carbon nanotube and polymer. Journal of Nanoparticle Research* 10 (8): 1309–1318.

107 Meng, Q., Cai, K., Du, Y. et al. (2019). *Preparation and thermoelectric properties of SWCNT/PEDOT:PSS coated tellurium nanorod composite films. Journal of Alloys and Compounds* 778: 163–169.

108 Cao, J.Q., Sun, Q., Miao, F.F. et al. (2014). *Preparation and thermoelectric power factor of Ag loaded carbon nanotubes/polyaniline composites. Materials Research Innovations* 18: 540–543.

109 An, C.J., Kang, Y.H., Lee, A.Y. et al. (2016). *Foldable thermoelectric materials: improvement of the thermoelectric performance of directly spun CNT webs by individual control of electrical and thermal conductivity. ACS Appl Mater Interfaces* 8 (34): 22142–22150.

7

Low-dimensional Thermoelectric Materials

Xinyi Chen, Yuanyuan Zheng, Xue Han, Yuanyuan Jing, Minzhi Du, Chunhong Lu, and Kun Zhang

Key Laboratory of Textile Science & Technology of Ministry of Education, College of Textiles, Donghua University, Shanghai, China

7.1 Introduction

Thermoelectric materials can be applied to power generation systems. Low-dimensional thermoelectric materials and their hybrids containing nanoparticles, nanowires/nanotubes/nanorods and nanofilms/nanosheets have promising thermoelectric performance due to their unique electronic structures. As more than one dimension in semiconducting materials are reduced to nanoscale, the electronic structure would be altered significantly, which is distinguishable from their bulks. In this chapter, we briefly review the thermoelectric performance of some typical low-dimensional thermoelectric materials, which may be helpful to understand their thermoelectric transports and contribute to develop high-performance macroscopic thermoelectric materials.

7.2 Zero-Dimensional (0D) Inorganic Semiconducting Nanocrystals

0D inorganic semiconducting nanocrystals mainly consist of IIB-VIB or IIIB-VB groups are generally referred to quantum dots (QDs). The size of QDs is in tens of nanometers in three dimensions. Within QDs, the

Flexible Thermoelectric Polymers and Systems, First Edition. Edited by Jianyong Ouyang.

electron transport is constrained, resulting in discrete energy levels similar to individual atoms [1]. As the size of semiconducting nanocrystals is reduced to match the exciton Bohr radius of bulk material, electrons are confined by the nanoscale boundaries, contributing to size-dependent properties [2].

The altered electronic structure will affect their thermoelectric performance. The Seebeck coefficient (S) is proportional to the difference between the average transport energy (E_{tr}) and Fermi energy (E_F). As the size of QD reduces, the band gap (E_g) increases and the band edge is pushed away from E_F, which results in the enhancement of S due to the enhancement of $E_{tr} - E_F$ [3]. For example, the S increases from 260 to $330\,\mu V\,K^{-1}$ at room temperature while the size of QD is reduced from 18 to 7 nm [4].

Meanwhile, in QDs with large E_g, it is difficult to activate electrons from valence band to conduction band, leading to limited carrier concentration (n) (and therefore deteriorated electrical conductivity). Wang et al. estimated that 8.6 nm nanocrystals had ~0.2 carriers/dot while 4.3 nm nanocrystals only possessed ~0.002 carriers/dot [3].

Furthermore, QDs have representative "phonon-blocking electron-transmitting" structure [5]. Liu et al. reported that the in-plane thermal conductivity (κ) of 50-period Ge/Si quantum dot superlattice (QDSL) was ~$30.5\ W\ m^{-1}\ K^{-1}$ at room temperature, which is much smaller than that of bulk Si [6]. The periodical sandwiched Ge quantum dot layers between Si layers strongly enhanced phonon scattering, while they did not severely impede electron transport because of the difference in mean free paths between electrons and phonons.

7.2.1 Measurements

Generally, the three thermoelectric parameters of 0D materials including the S, electrical, and thermal conductivity can be measured by specially designed setups. For example, the S and electrical conductivity (σ) of QD films are measured in air using a microscopic device comprising of a microheater and a thermometer (Figure 7.1a). These microscopic devices are designed to measure low-dimensional materials with high spatial resolution. A temperature gradient ΔT is applied on the QD film by Joule-heated microheater. The local temperatures T_1 and T_2 are measured by characterizing temperature coefficient of resistances. The κ of the QD film can be measured via 3ω method [7] (Figure 7.1b) under ultrahigh vacuum conditions. A Gwinstek frequency synthesizer is used to generate a 10–1000 Hz

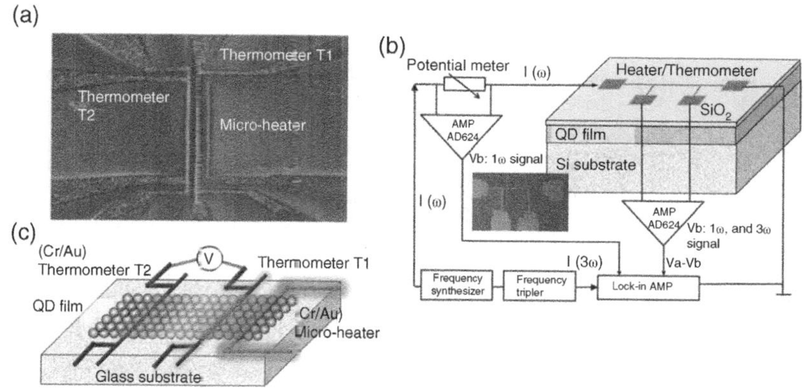

Figure 7.1 (a) The SEM image of the *S* measurement patterns with false-colored metal lines, and (b) the schematic illustration of micro-device for *S* measurement. (c) Schematic diagram of the 3ω thermal conductivity measurement apparatus and a SEM image of measurement pattern (inset). *Source:* Dajiang et al. [4]. © Royal Society of Chemistry

sine wave, and a lock-in amplifier is used to extract 3ω voltage signals. The κ of the QD film is calculated by [4]:

$$k_{\mathrm{f}} = \frac{pd_{\mathrm{f}}}{2b\left\{\Delta T_{\mathrm{s+f}}(\omega) - \Delta T_{\mathrm{s}}(\omega)\right\}} \tag{7.1}$$

where $\Delta T_{\mathrm{s+f}}(\omega)$ and $\Delta T_{\mathrm{s}}(\omega)$ are temperature oscillations from the substrate with and without the thin film, respectively. The parameters *p, b,* and d_{f} are the unit heating power, the half width of heater and thickness of the QD film, respectively.

7.2.2 Materials and Properties

7.2.2.1 Fullerene

Fullerene (C_{60}), an allotrope of carbon, possesses a spherical bulkyball shape [8] (Figure 7.2b). The carrier density of the intrinsic C_{60} film is very low leading to the low σ [11]. C_{60} molecule is a face-centered cubic (*fcc*) structure when the temperature is higher than 260 K. At 260 K, C_{60} undergoes a first-order transition from *fcc* to simple cubic (*sc*) structure accompanied by a reduction of orientation, leading to a high κ ($>1\ \mathrm{W\,m^{-1}\,K^{-1}}$) [12, 13].

It is well known that C_{60} has a negative *S*. However, Almutlaq et al. [9] studied thermoelectric properties of the single molecule junction of the $C_{50}Cl_{10}$ with exohedral Cl atoms and found that it showed a positive *S*

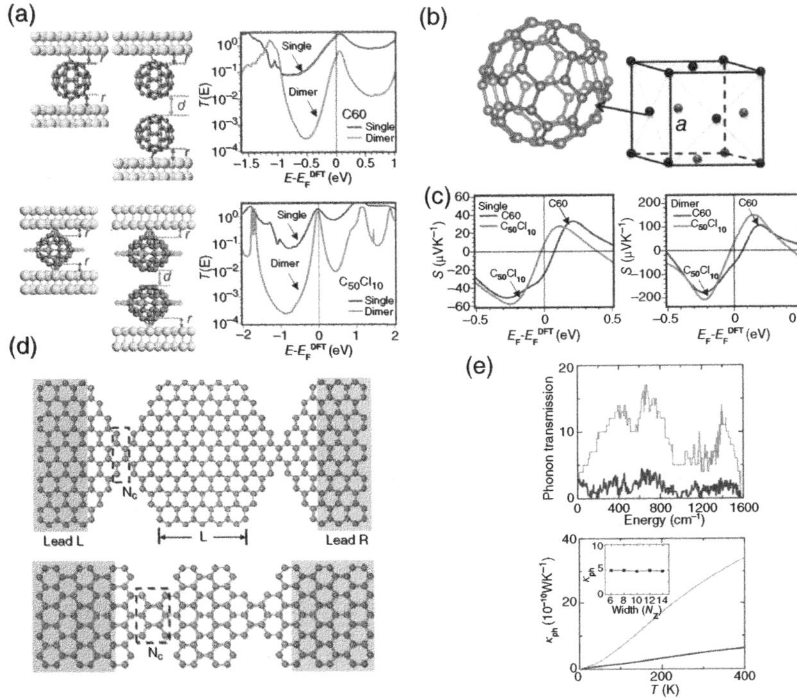

Figure 7.2 (a) A single C_{60} (top figure) and a C_{60} dimer (bottom figure) placed between two gold electrodes. *Source:* Almutlaq et al. [9]. © Royal Society of Chemistry. (b) Molecular structures of C_{60}. *Source:* Chen et al. [8]. © Springer Nature. (c) The calculated *S* depends on the difference between Fermi energy E_F relative and DFT-predicted Fermi energy E_F^{DFT}. *Source:* Almutlaq et al. [9]. © Royal Society of Chemistry. (d) A nanoribbon with zigzag edges and armchair edges. (e) Phonon transmission for the ZGNR with the narrowest constrictions (thin line) and phonon transmission of the pristine ZGNR (thick line). Phonon thermal conductance of the nanoribbon with the narrowest constrictions (thin line) and the pristine ZGNR (thick line). The inset shows the thermal conductance of ZGNRs with the narrowest constrictions when the width is varied at $T = 300$ K. thermal conductance for ZGNRs and AGNR structures, respectively [10]. *Source:* Yan et al. [10]. © Elsevier.

(Figure 7.2a). Figure 7.2c shows that the *S* of monomers depends on the energy difference between the Fermi energies (E_F) and the DFT-predicted Fermi energy (E_F^{DFT}). The change in the sign of *S* can be ascribed to the E_F of $C_{50}Cl_{10}$ above the highest occupied molecular orbital (HOMO) level, while E_F of C_{60} is below the lowest occupied molecular orbital (LUMO) level. The *S* of $C_{50}Cl_{10}$ monomer and dimer are 8 and 30 μV K^{-1}, respectively. The reason for this increase is that the value of transmission coefficient ($T(E)$) near the LUMO level remains close to unity for both the monomer and dimer whereas the slope of the transmission coefficient is larger for the dimer (Figure 7.2a, Formula (7.2) [14]).

$$S(T) \approx -\alpha e T \left(\frac{d \ln(T(E))}{dE} \right)_{E=E_F} \tag{7.2}$$

where $\alpha = \left(\frac{k_B}{e}\right)^2 \pi^2/3$ is the Lorenz number, k_B is Boltzmann's constant, and e is the electronic charge, Eq. (7.2) demonstrates that S is enhanced by increasing the slope on $\ln T(E)$ near E_F.

7.2.2.2 Graphene Quantum Dots

Graphene quantum dots (GQDs) with a lateral size less than 20 nm, are emerging materials for numerous applications due to their appealing characteristics, such as high specific surface area, good biocompatibility, low toxicity, and the available π–π conjugated network with functional surface groups [15–17].

Yan et al. [10] investigated thermoelectric properties of graphene nanoribbons (GNRs) in two constricted states. The central part between the two constrictions can also be viewed as a hexagonal GQD (Figure 7.2d). This structure can be obtained by engraving narrow constrictions on pristine GNRs. Figure 7.2e shows the phonon transport of the zigzag GNR (ZGNR) with the narrowest phonon constriction (thick line) and phonon transmission of the pristine ZGNR (thin line). The transmission of the pristine ZGNR is staircase-like, corresponding to the phonon channel at a given energy. Long-wavelength phonons can pass through objects with sizes smaller than the phonon mean free path, while short-wavelength and medium-wavelength phonons would be selectively scattered. The directional flow of phonons means the energy transport and heat conduction. So, in the case of ZGNRs with constrictions, the phonon transport with medium or short wavelength is largely suppressed by constrictions. Hence, the thermal conductance of ZGNR with constrictions is smaller due to strong phonon scattering. It has been reported that ZGNRs with constrictions show a reduced thermal conductance of ~0.5 nW K^{-1}, which is much lower than the pristine ZGNRs (2.6 nW K^{-1}, Figure 7.2e) at 300 K.

7.3 One-Dimensional (1D) Thermoelectric Materials

1D materials are a kind of materials with diameters in nanoscale including nanotubes, nanorods (aspect ratios greater than 1 but less than 20), and nanowires (aspect ratios greater than 20) [18]. Generally, the synthetic methods of 1D organic materials include electrodeposition [19, 20], self-assembly using nanopore anodic aluminum oxide (AAO) templates [21, 22], electrospinning [23–25], and other techniques [26, 27]. At present, 1D thermoelectric materials with high performance have become hot topics.

7.3.1 1D Organic Thermoelectric Materials

Compared with organic bulk materials, 1D conducting polymers usually exhibit excellent thermoelectric properties [28]. Meanwhile, 1D conducting polymers maintain the outstanding flexibility, which can realize flexible energy generation or heating–cooling devices with high performance and conformal geometries. In this section, we will focus on conducting polymer nanowires to discuss their thermoelectric properties. And the chemical structures and properties of 1D conducting polymers have been shown in Figure 7.3 and Table 7.1, respectively.

7.3.1.1 Poly(3,4-Ethylenedioxythiophene) Nanowires

Of particular interest are poly(3,4-ethylenedioxythiophene) (PEDOT) with remarkably high conductivities, outperforming all other conducting

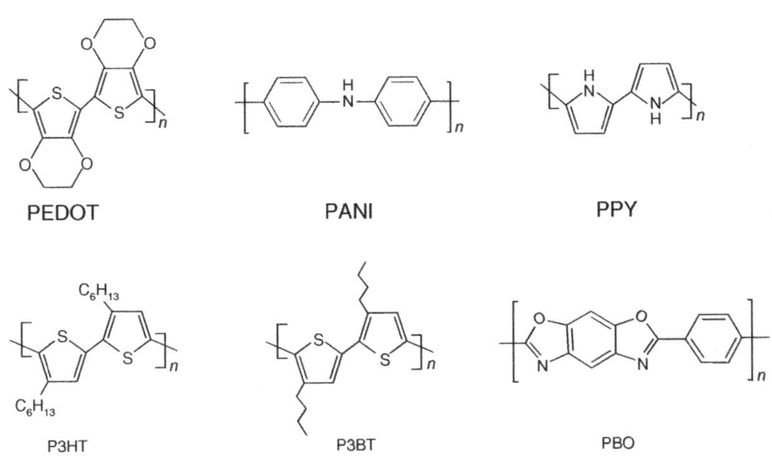

Figure 7.3 Typical chemical structures of conducting polymer nanowires. *Source:* Xinyi Chen.

Table 7.1 Experimentally measured σ, S, PF, μ and k values for 1D conducting polymers reported in the literature.

Sample	σ (S cm^{-1})	S (µV K^{-1})	PF (µW m^{-1} K^{-2})	μ (cm^2 V^{-1} s^{-1})	k (W m^{-1} K^{-1})
PEDOT:Cl nanowires [26, 29]	~140–541	~20.78–50.55	~23.4–35.8	~4.5–6.62	~0.25
PEDOT:Tos nanowires [26]	~236	~20.78	~10.3	~1	
Intrinsic PEDOT nanowires [30]	~6.6–39.6	~33–80	~$(0.96–8.4)\times10^{-6}$	~2.1–13.2	—
Single-crystal PEDOT nanowires [31]	~7619–8797	—	—	~88.1	—
PEDOT nanotube [32]	~6.4×10^{-5}	~4088	~1.07	—	—
PANI nanowires [33]	~$(1–100)\times10^{-1}$	~5.5–32.1	~0.07–2.7	—	~0.22–0.42
PANI nanorods [33]	~1.25	~5	~1.2×10^{-3}	—	~0.39
PPy nanowires [34, 35]	~$(3–150)\times10^{-1}$	~6.8–9.8	~$(1.5–120)\times10^{-3}$	—	—
PPy nanotubes [35]	~10.4	~6.0	~0.32	—	—
P3HT nanowires [28, 36]	—	—	—	~0.06	~0.5–2.29
P3BT:PS nanowires [37]	~10^{-3}–10^{-2}	~525–600	~$(0.8–7.3)\times10^{-2}$	~0.01	~0.12–0.33
PBO nanowires [38]	—	—	—	—	~20

Sources: Based on Refs. [26, 28–38]. © John Wiley & Sons.

polymer [39]. For organic semiconductors, their thermoelectric performance has a close relationship with the crystallinity, orientation degree, chemical structure, and electronic structure. Generally, S and σ depend on the n via the E_F whereas they have different trends – increasing n fills more states and dramatically boosts σ but also brings E_F closer to them, decreasing S [40]. PEDOT nanowires (NWs) were reported to display sharp density of state (DOS) at E_F [26]. A larger slope of DOS at E_F can push E_F away from densely spaced states and contribute to improve the average energy difference per carriers ($E - E_F$), leading to higher S [41]. Regarding the σ, PEDOT NWs with crystalline characteristics have high carrier mobility (~4.5 $cm^{-2}\ V^{-1}\ s^{-1}$), which contributes to higher σ [26, 29].

Main strategies of synthesizing PEDOT NWs include soft-template [33, 34], hard-template [22, 36, 42], and interfacial polymerization [39, 43, 44] method. Since it is difficult to accurately control the nanowire morphology, the interfacial polymerization method is rarely used. The soft-template method is the most common strategy to synthesize PEDOT NWs [45]. It utilizes high surfactant concentrations to form micelle templates with hydrophilic and hydrophobic groups. Due to the hydrophobic nature of EDOT monomers, they spontaneously enter the interior of micelle and then are oxidized by oxidant. The resultant PEDOT NWs possess a diameter of ~10 nm and a length of ~1.5 μm. The σ of PEDOT NWs can achieve up to 1340 S cm^{-1} [46]. And the power factor (*PF*) of PEDOT NWs is 10.3–35.5 μW $m^{-1}\ K^{-2}$ [26].

The hard-template method is an important way to precisely control the geometry of nanowires [30]. The lithographically patterned nanowire electrodeposition (LPNE) technology was used to prepare PEDOT NWs in an aqueous $LiClO_4$ solution, by which the length of PEDOT NWs was up to 200 μm. The thermoelectric performance of a single "orphaned" PEDOT NW plus an array of PEDOT NWs were measured by specially designed micro-devices. PEDOT NWs were patterned onto a Si_3N_4-coated Si wafer. The thermal voltage and temperature difference between two ends of nanowires were measured by two thermocouples within a vacuum cryostat. At 310 K, the σ and S of a single PEDOT NW were measured to be 6.9–40.5 S cm^{-1} and 33–122 μV K^{-1}, respectively. Its *PF* reached to ~9.2 μW $m^{-1}\ K^{-2}$, which is higher than that of PEDOT films with an identical electrodeposition process (~2.6 μW $m^{-1}\ K^{-2}$). The higher S of PEDOT NWs can be attributed to lower n than that of PEDOT films. As aforementioned, the lower n leads to lower σ. However, due to the different orientational orders of molecular chains and the crystallinity of PEDOT NWs, PEDOT NWs possess higher carrier mobility (~9 $cm^2\ V^{-1}\ s^{-1}$) than films (~3 $cm^2\ V^{-1}\ s^{-1}$), leading to excellent σ.

Nowadays, single-crystal PEDOT NWs with ultrahigh conductivity have been directly developed by liquid-bridge-meditated nanotransfer printing with vapor phase polymerization [31]. In this method, EDOT monomers were oxidized, self-assembled, and crystallized in nanoscale channels of the template containing the $FeCl_3$ oxidant. This single-crystal PEDOT NWs had smaller lateral chain spacing and π–π stacking distance than previous attempts, perhaps because of the relatively small dopant (Cl^-) and compact packing of the EDOT monomers within the nanoscale channels. This molecular structure contributed to high carrier mobility of 88.08 $cm^{-2}\ V^{-1}\ s^{-1}$, leading to an ultrahigh $\sigma \sim 8797\ S\ cm^{-1}$.

7.3.1.2 Other Polymer Nanowires

Similar to PEDOT, polyaniline (PANI) is a kind of common conducting polymer, showing good σ, flexibility, stability, and processability. Changing doping level is an effective way to optimize thermoelectric performance for PANI. During doping process, the n of conducting polymers will be improved, contributing to the relatively high σ and low S. The peak value of power factor ($PF = S^2\sigma$) can be achieved by the trade-off between S or σ. For PANI polymer, the doping level can be changed by protonation process. The alkaline imine and amine nitrogen atom in PANI polymer are easier to be protonated under acidic condition. Electrons will be grabbed from PANI chains during the protonation process, leading to p-type doping for PANI chains. Generally, intrinsic PANI is similar to an insulator ($<10^{-9}\ S\ cm^{-1}$) [47]. However, the σ of PANI NWs was sharply increased from $10^{-1}\ S\ cm^{-1}$ to $10\ S\ cm^{-1}$ under different acid concentrations [33]. And the S for PANI NWs decreased from 31.5 to 5.2 $\mu V\ K^{-1}$ [33]. The κ of PANI NWs can be improved by increasing doping level of the semiconductor with low doping level because the enhancement of n contributes to heat conveyance. Interestingly, the κ will be reduced for the PANI NWs with high doping level, due to the enhancement of ion scattering [48]. So the κ of PANI NWs with different doping level ranged from 0.2 to $0.4\ W\ m^{-1}\ K^{-1}$ [33]. Different kinds of acids also affect the performance of PANI NWs. The performance of PANI NWs doped by p-toluenesulfonic acid ($ZT = 2.75 \times 10^{-5}$) was four times than that of samples doped by hydrochloric acid, which was probably due to the large volume of the p-toluene sulfonic acid anion contributing to the ordered arrangement of PANI segments [49].

Polypyrrole (PPY) NWs are usually synthesized by chemical polymerization with oxidants including KIO_3, $KMnO_4$, K_2CrO_4, $KBrO_3$, $KClO_3$, $(NH_4)_2S_2O_8$ (APS), etc. [50]. Compared with other 1D conducting polymers, reports about PPY NWs are limited. The chemical structure of PPy NWs is similar to PANI NWs. So protonation process is also a useful way to change its thermoelectric performance [51]. And it has been demonstrated that the morphology of nanowires will affect their performance [34]. The morphology of PPY NWs synthesized by APS oxidant was curly. With the polymerization time increasing, PPY NWs tended to be straight and thick [34]. And the straight PPY NWs have been proved to display better thermoelectric performance ~$(22.6 \pm 3.6) \times 10^{-3}$ μW m^{-1} K^{-2}). The reason was that straight NWs probably had better crystalline leading to high carrier mobility.

Recently, poly(3-hexylthiophene) (P3HT) NWs have gained a lot of attention in the field of organic field effect transistors. Since the carrier mobility of a single P3HT NW in field-effect transistors can be as high as 0.06 cm^2 V^{-1} s^{-1} [52]. Meanwhile, P3HT NWs have a strong affinity for organic solvents and high molecular weight controllability, making it easy to be large-scale processed into P3HT NWs [53–55]. As the molecular weight (Mn) of P3HT is higher than a critical molecular weight (number-average Mn ~10 kDa), the regular stereo-structured P3HT can be self-assembled into a highly crystalline nanowire by π–π conjugation without assistance of templates [56, 57]. In addition to the above synthetic method, P3HT NWs also can be produced by hard-template method [22, 58]. The κ of an individual P3HT NW synthesized by hard-template method can be measured by a technique called 3ω-SThM (scanning thermal microscopy) based on scanning probe microscopy (SPM). This technique could capture thermal images of independent NW by a 100 nm representative thermal spatial resolution and a representative 10 nm topographical spatial resolution [36]. And it could probe a large range of individual NWs within its matrix in a short period of time. The κ of P3HT NWs were from 2.29 ± 0.15 to 0.50 ± 0.24 W K^{-1} m^{-1}, as the diameter of NWs reduced from 350 to 120 nm. This phenomenon was attributed to an increase of crystalline-oriented lying with π–π stacking direction parallel to the nanowire long axis in narrow nanowires.

Poly(3-butylthiophene) (P3BT) NWs have the high degree of crystallinity lying side-chain direction (or namely <010>) [37]. P3BT molecules can crystallize and self-organize to form P3BT NWs without template [59]. In P3BT NWs, hole concentration is 10^{24}–10^{25}. And they have the higher S (~900 μV K^{-1}) and lower σ (~10^{-4} S cm^{-1}). To improve σ without sacrificing S, P3BT NWs are usually blended with insulating polystyrene (PS). Because of immiscibility between conjugated polymers and insulating polymers, their physical blend usually leads to large-scale phase separation, contributing to interpenetration conductive network within insulating matrix [59]. The σ of P3BT/PS (40–60% P3BT) was increased to ~10^{-3}–10^{-2} S cm^{-1}.

Polybenzobisoxazole (PBO) NWs are extremely special NWs because of their high thermal conductivity [60]. According to previous research, liquid crystalline PBO nanowires had the high κ of ~20 W m^{-1} K^{-1} [38]. Because group velocities of longitudinal acoustic phonons along nanowires length play dominant roles in heat conduction for PBO nanowires with the high degree of elastic anisotropy.

7.3.2 Carbon Nanotubes

Carbon nanotubes were discovered by Iijima in 1991 [61], which can be divided into single-walled carbon nanotubes (SWCNTs) and multi-walled carbon nanotubes (MWCNTs) based on the number of walls [62]. It is theoretically possible to construct a sp^2-hybridized carbon nanotube by rolling up a hexagonal graphene sheet and thus leading to "non-chiral" (armchair and zigzag configuration) and chiral arrangements (Figure 7.4) [63]. In the non-chiral configuration, the honeycomb lattices, located at the top and

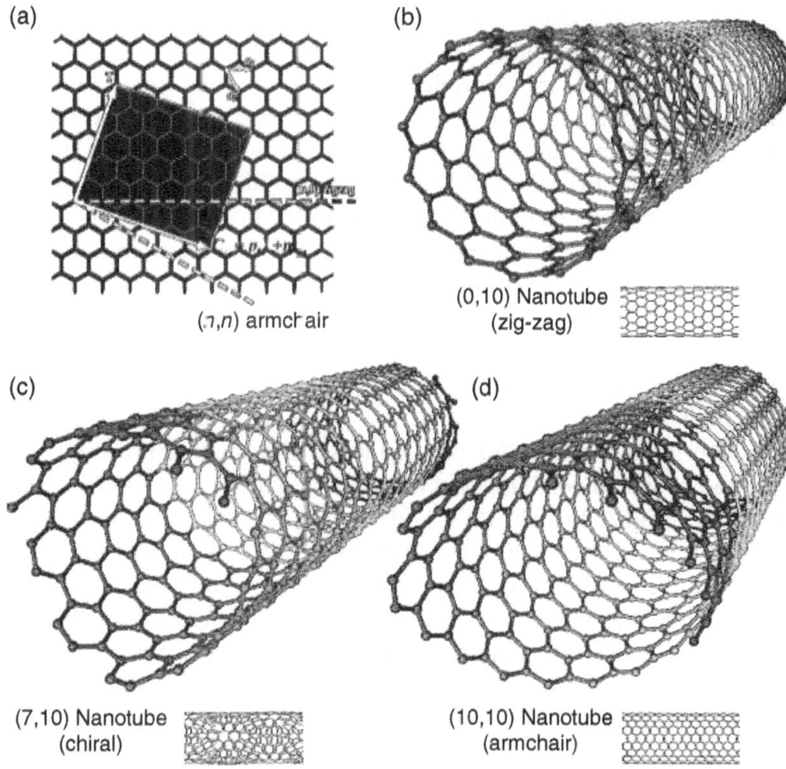

Figure 7.4 (a) Molecular models of exhibited by SWNTs based on the chirality: (b) zigzag, (c) chiral, and (d) armchair based conformation [63]. *Source:* Mittal et al. [63]. © Elsevier.

bottom of the tube, are always parallel to the tube axis (these configurations are known as armchair and zig-zag). In the armchair structure, two C–C bonds on opposite sides of each hexagon are perpendicular to the tube axis, whereas in the zig-zag arrangement, these bonds are parallel to the tube axis (Figure 7.4b and d). All other conformations in which the C–C bonds lie at an angle to the tube axis are known as chiral or helical structures (Figure 7.4c). Depending on their different electronic structures along with chirality, SWCNTs can be metallic (m-) or semiconducting (s-) [64, 65] and have diameter-dependent band structures [66–68]. The s-SWNTs exhibit a significant band gap near the E_F, while m-SWNT shows overlapped conduction and valence bands. Among synthesized SWNTs, the typical mass ratio between s-SWNT and m-SWNT is ~2 : 1 [69]. However, MWNT has unique electronic properties, and both metallic and semiconducting behaviors are observed. Both the metallic and semiconducting nature of the tubes are expected to be modified by the interlayer interactions [70].

Electronic structures of CNTs can be experimentally studied by scanning tunneling microscopy (STM). They can also be investigated by theoretical calculation using the first-principle calculation. The band gap of SWCNTs was reported in the range of 0.2 to 1.2 eV [71]. Gao et al. reported that a small fraction of nanotubes has a work function of ~5.6 eV, which is ~0.6 eV higher than that of carbon [72]. In ambient condition, CNTs show p-type characteristics, which is originally because of the doping effect by oxygen and moisture in air [73]. As oxygen was gradually removed from CNTs, the S gradually changed from positive to negative. When oxygen was reintroduced into CNTs, the S became positive again which indicates a n-type behavior of intrinsic CNTs [74, 75]. The stable and perfect sp^2 structure of CNTs contributes to the high carrier mobility in CNTs [76–78], resulting in a very high σ and κ.

The σ of an individual CNT can be measured using four-probe method and two-probe method. Ebbesen et al. used the four-probe method to measure the σ of an individual CNT [70]. The measured σ of an individual MWCNT can reach 1.96×10^5 S cm^{-1}. This method can diminish the contact resistance and thus improve the measurement accuracy. There is another simple two-probe method to measure the σ of an individual CNT [79–81]. This method is relatively easy to be carried out, but the contact resistance between CNTs and electrodes cannot be eliminated. Therefore, the measured σ is smaller than its real values.

In terms of thermal transport, CNTs possess very high thermal conductivity [82, 83]. The thermal conductivity measurement involves "thermal conductance" technology by specially designed micro-bridges [84–87]. This method can simultaneously obtain κ and S of individual CNT. The measured

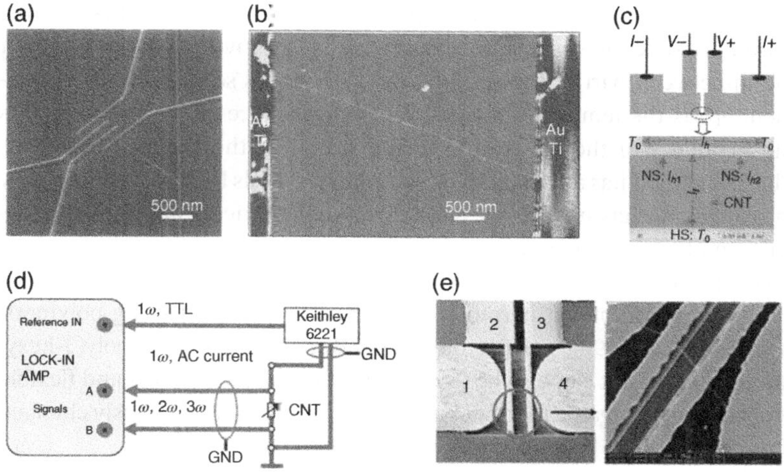

Figure 7.5 Different methods for measuring σ, S, and the κ of individual CNT. (a and b) σ measurement. (a) Four-probe method and (b) two-probe method. *Sources:* (a) Lezec et al. [83] and (b) Kong et al. [79]. © with permission from the Springer Nature. (c and e) S and σ measurement. (c) T-type nanosensor measurement. *Source:* Fujii et al. [82]. © American Physical Society. (d) Schematic diagram of 3ω measurement setup and (e) four-wire construction by focused ion beam milling (left); selective deposition of single CNT by dielectrophoresis (right). *Sources:* Choi et al. [88]. © American Chemical Society.

κ and S of MWCNT are ~3000 W m^{-1} K^{-1} and ~80 μV K^{-1}, respectively. The measured κ and S of SWNT are ~2000 W m^{-1} K^{-1} and ~42 μV K^{-1}. Another method for κ measurement involves the usage of a T-type nanosensor. Fujii et al. [82] developed this method in 2005 (Figure 7.5c). The measured κ is ~2069 W m^{-1} K^{-1} for an individual MWCNT with a diameter of 9.8 nm at room temperature. Besides these methods, the 3ω method is also very common for measuring thermal conductivities of CNTs (Figure 7.5d and e) [88–90]. In such methods, by loading an alternating current in the form of $I_0\sin\omega t$ into the two outside pads, the temperature rise of suspended specimen fluctuates at an angular frequency of 2ω due to the Joule effect. The angular frequency further results in a 3ω voltage harmonics, $U_{3\omega}$, which can be used to deduce the axial thermal conductivity of film. It is given by [90]:

$$k = \frac{4U_{1\omega}{}^{3}L\alpha_{CR}}{\pi^4 U_{3\omega} R_0 S\sqrt{1+\tan\varphi^2}} \tag{7.3}$$

$$\tan\varphi = \frac{4fL^2}{\pi\alpha} \tag{7.4}$$

where R_0, $U_1\omega$, and $U_3\omega$ are the resistance, the amplitude of the first and the third harmonics across the specimen heater, respectively; L is the length of the in-between part of the specimen, S is the cross sectional of the specimen, α_{CR} is the temperature coefficient of resistance of the specimen, f is the frequency of the alternating current, and α is the thermal diffusivity. The 3ω method has an advantage over other methods because it is very easy to realize and has no contact resistance problem, negligible spurious signals, and low equipment cost.

Due to the σ and mechanical properties, CNTs have been widely used as nanofillers to improve thermoelectric properties of conducting polymers, such as PEDOT [91], polypyrrole [92], polyaniline [93], and poly(3-hexyl thiophene) [94]. It can also be used to improve the mechanical and flexible properties of inorganic thermoelectric materials without obviously changed figure of merit [95].

7.4 Two-Dimensional (2D) Thermoelectric Materials

2D materials, also known as atomic-thin films, usually are several nanometers or even sub-nanometers thick (e.g. graphene) with a few atom layers in thickness direction. 2D inorganic materials can be obtained from layered materials (e.g. graphite and phosphorus) by mechanical and liquid exfoliation or chemical vapor deposition [96–98]. 2D organic materials, including small organic molecules and polymer nanofilms or nanosheets can be prepared using self-assembly strategy [99–101], molecular layer epitaxy [102, 103], chemical or physical exfoliation of layered polymer crystals [104–107], surface-catalyzed polymerization, etc. [108–111]. The electronic structure and physicochemical properties of 2D materials are distinctive from those of their bulks. As discussed in the previous sections, the motion of electrons in two-dimensional materials are restricted in the in-plane direction. Moreover, 2D nanocrystals may emerge the so-called quantum confinement effect as the thickness is comparable to the Bohr radius of carriers which will discrete energy levels of electrons and split density of states into sub-bands [112, 113]. The unique electronic structures make 2D materials have great potentials in optoelectronics, thermoelectrics and supercapacitors. In this part, we will review some typical 2D materials with potentials for thermoelectric applications including graphene, conducting polymer nanosheets or nanofilms, and some notable inorganic 2D materials (e.g. black phosphorus and MXene).

7.4.1 Graphene

Graphene is a single atomic carbon layer, which was discovered by Andre Geim and Konstatin Novoselov through mechanical exfoliation of bulk graphite in 2004 [114]. This discovery opens the era of 2D materials. Theoretically, graphene can be considered as a 2D building block (Figure 7.6) for carbon nanocrystals in all other dimensionalities [96], including fullerenes (0D), carbon nanotubes (1D), and graphite (3D). Graphene is ~0.35 nm in thickness which has been recognized as the thinnest material to date. Similar to graphite, carbon atoms in graphene form hexagonal honeycomb lattice by sp^2 hybrid orbital arrangement [97].

Single-layer graphene is a zero-bandgap semimetal meaning that the valence band touches the conduction band at a single Dirac point [96, 115] (Figure 7.6b). The electron mobility of graphene is as high as $\sim 2 \times 10^5\ cm^2\ V^{-1}\ s^{-1}$ [116, 117] which is two orders of magnitude higher than that of silicon [118]. It has been widely used in almost all areas of electronics such as thermoelectrics, rechargeable batteries, supercapacitors, sensors, etc. However, intrinsic graphene apparently is not a good candidate to thermoelectrics because of the ultra-high κ (i.e. $\kappa \sim 1600–1800\ W\ m^{-1}\ K^{-1}$ at $T = 300$ K) [119]. The zero-bandgap semi-metallic properties of graphene limits the S ($10^0–10^1\ \mu V\ K^{-1}$) as well [120].

The S of the monolayered graphene can be measured using microfabricated devices. Local temperature variations are measured with two metal four-probe micro-thermometers (Figure 7.7a) [121]. The S of 39 $\mu V\ K^{-1}$ was measured in the monolayered graphene at 255 K [122]. The spatially

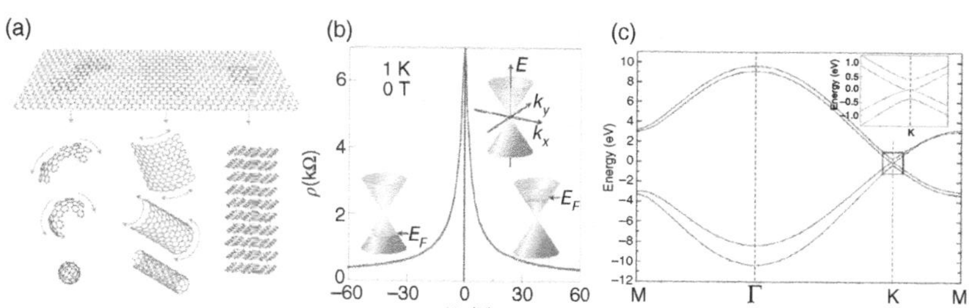

Figure 7.6 (a) Different carbon allotropes that can be created by using graphene as the basic unit. (b) Ambipolar electric field effect in a single-layer graphene. The insets show its conical low-energy spectrum $E(k)$, indicating changes in the position of the Fermi energy E_F as the gate voltage V_g changes. (c) The band structure of graphene bilayers along $M\Gamma KM$. The square region is enlarged in the inset. *Source:* Geim et al. [96]. © Springer Nature.

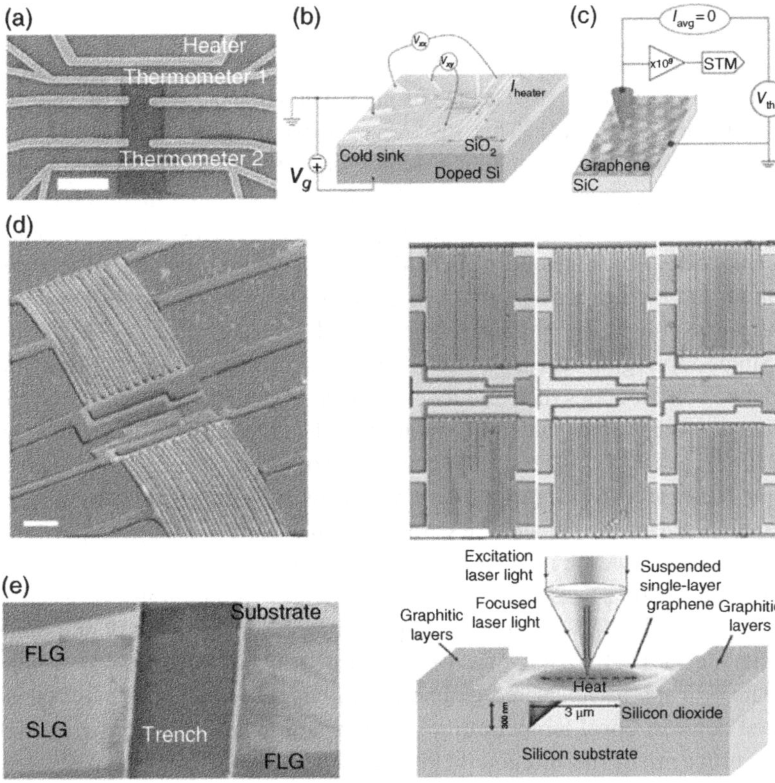

Figure 7.7 (a–c) Different measurement methods for S of graphene. (a) Micro-device using conventional method; (b) circuit schematic of the graphene device for S measurement. *Sources:* Zuev et al. [121] and Wei et al. [122]. © American Physical Society; (c) simultaneous structure and thermovoltage measurement with an STM at the atomic resolution on epitaxial graphene on SiC. *Source:* Park et al. [123]. © American Chemical Society. (d–e) Different methods of measuring κ of graphene monolayer. (d) electro-thermal micro-bridge method. *Source:* Xu et al. [124]. © Springer Nature. (e) the SEM image of the suspended graphene across a trench in Si/SiO_2 substrate (left) and schematic of the experiment showing the excitation laser light focused on a graphene layer suspended across a trench (right). *Source:* Balandin et al. [125]. © American Chemical Society.

resolved S of monolayered graphene on SiC can be mapped using STM method [123]. The S can be calculated by measuring the thermovoltage between the tunneling tip and sample caused by a ΔT across the tunneling barrier of a scanning tunneling microscope. The S of monolayer graphene was measured to be 42 $\mu V\ K^{-1}$ using this method. Besides the measurement of S, the STM system equipped with high-resolution transmission electron microscopy (HRTEM) can also be used to measure the σ of single layer graphene [126]. The graphene powders were attached to an Au electrode, and the tungsten STM tip was controlled precisely in the HRTEM to connect with the graphene to measure its σ. The σ of monolayer graphene obtained from artificial graphite is 1000 S cm^{-1124}. Li et al. found an interesting phenomenon that the S of multilayered graphene is layer-dependent. It increases with layer number and reaches a peak value of 54 μV K^{-1} at six layers by using the method presented in Figure 7.7a. Meanwhile, the σ monotonically decreases as the layer number increases [127]. The layer-dependence of S is peculiar and unexpected because the S of graphene is determined by the interplay of the band structure, the defects or disorder induced scattering, the temperature and the Landau quantization.

The micro-bridge technology with two suspended micro-heaters (Figure 7.7d) [124] firstly reported by Okuda in 1992 [128] has been frequently used to measure the κ of graphene. As reported by Xu et al., the κ of graphene was measured to be ~1689–1813 W m^{-1} K^{-1} at $T = 300$ K (length = 9 μm) [129]. At 300 K, the κ was found to scale with its length (L) as $\kappa \sim \log L$ even when L is one order of magnitude larger than the average phonon mean free path. Xu et al. analyzed that this length-dependent κ is related to the 2D nature of phonons in graphene which indicates the fact that phonons in graphene can travel without scattering. This is different from those in three-dimensional or one-dimensional systems. In 3D bulk materials, heat conduction is governed by the Fourier's law: $J = -\kappa \Delta T$, where J is the local heat flux, ΔT is the temperature gradient and κ is thermal conductivity. κ is an intrinsic parameter of the materials and is usually size and geometry independent. However, for 1D lattice models, quasi-1D nanostructures and polymers, a length dependence of κ as $\sim L^{\beta}$ has been predicted theoretically and then experimentally verified for nanotubes. Because graphene has clear signatures in Raman spectra and the G band in graphene spectra manifest a strong temperature dependence [130–132]. There is a noncontact thermal conductivity measurement for single layered graphene using the confocal micro-Raman spectroscopy [125, 131] (Figure 7.7e). The thermal conductivity can be calculated by $\kappa = \chi_G (L/2hW)(\delta\omega/\delta P)^{-1}$, where χ_G is the first-order temperature coefficient, L is the distance from the middle of the suspended graphene to the heat sink with the ambient temperature, h is the height of graphene, W is the width of graphene, $\delta\omega$ is the a small shift in the G band peak position due to the variation δP (laser excitation power change) in the heating power on the sample surface. Compared to the suspended micro-bridge method, this method is much easier and has no damages to graphene. The local temperature in graphene rises as a function of applied laser power which then is utilized to extract the thermal conductivity. The measured κ of single layer graphene is as high as 1800 W m^{-1} K^{-1} at a temperature of 325 K [131].

7.4.2 Black Phosphorus

Black phosphorus (BP) is the most thermally stable one among phosphorus allotropes [133]. BP possesses layered structure [134–136] in which vertically aligned atomic BP layers stack together via Van der Waals force [98, 137–140] (Figure 7.8). Similar to graphene, 2D BP monolayer can be fabricated with mechanical or liquid exfoliation, electrochemical method, plasma etching, chemical vapor deposition, etc. [133, 138, 141]. In the individual BP monolayer, phosphorus atoms are covalently bonded with three adjacent phosphorus atoms through sp^3 hybridization [142].

Intrinsic defects and impurities are usually unavoidable in BP. Both point vacancies and self-interstitials play the prevailing role of shallow acceptors, therefore BP is typically observed as a p-type semiconductor [143]. BP has a direct band gap of ~0.34 eV in bulk and ~2.0 eV in monolayer [144]. This is a very appealing feature as the band gap value of BP spans over a wider energy range than other 2D materials (Figure 7.9f) [145]. The band gap, band alignment, work function, band structure, and effective carrier mass

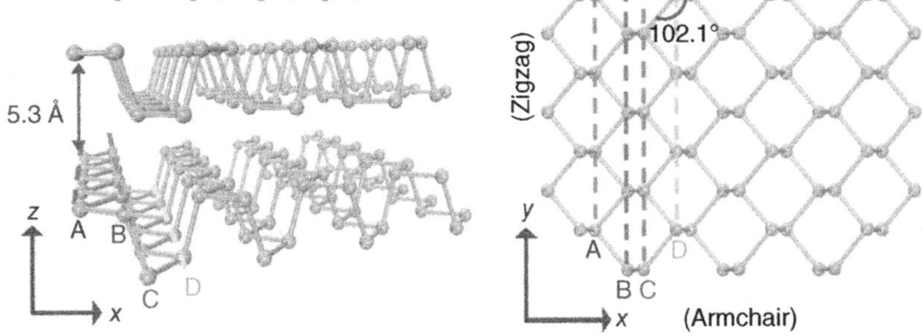

Figure 7.8 (a) Atomic structure of BP. It consists of puckered phosphorus atom planes with strong intralayer sp^3-hybridized covalent bonds and weak van der Waals interlayer interaction with the interlayer space of 0.53 nm. (b) The top view of the lattice of monolayer BP. The bond angles are shown. The x and y directions correspond to the AC and ZZ directions of BP, respectively. *Source:* Ling et al. [135]. © National Academy of Sciences

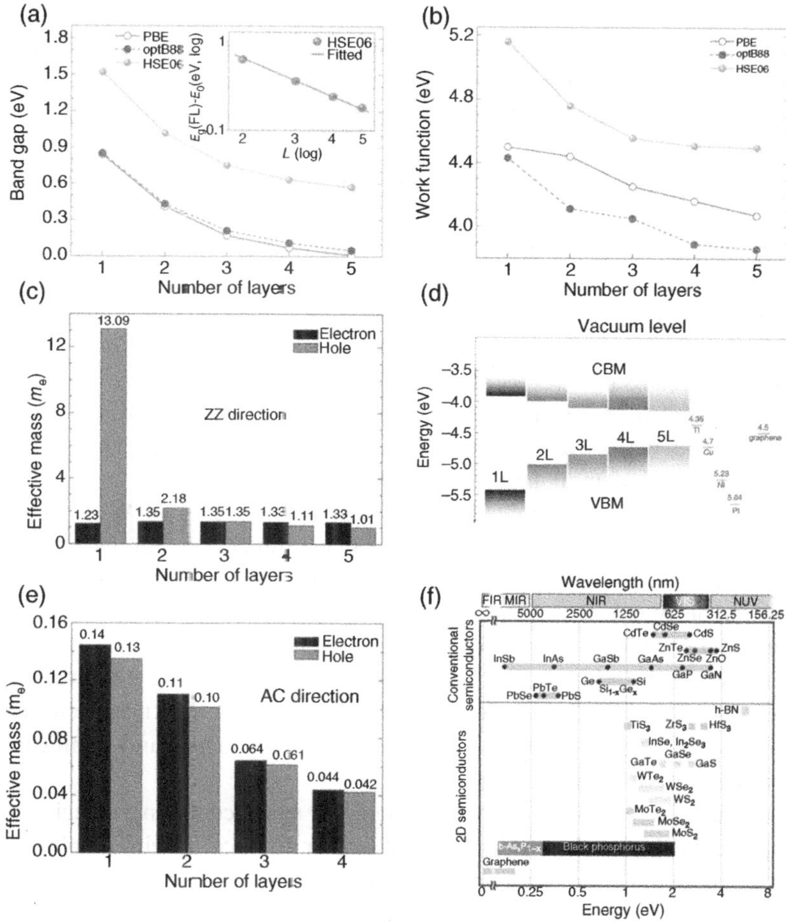

Figure 7.9 (a–e) The layer-dependent band gap (a), work function (b), effective mass (c and e), and band structure (d) of BP. *Source:* Cai et al. [136]. © Springer Nature. (f) The summarized band gaps of other 2D semiconductors compared to that of BP. *Source:* Castellanos-Gomez [145]. © with permission from the American chemical Society.

are all layer-dependent (Figure 7.9a–e) [146]. Few-layer BP has smaller carrier effective mass, higher carrier density, and weaker carrier scattering due to enhanced electrostatic interlayer screening effect when compared with monolayer BP. The layer-dependent band edges and work functions of few-layer phosphorene allow the facile modification of Schottky barrier to achieve enhanced carrier injection efficiency.

The lattice structure of BP determines the unique characteristics of anisotropy in the BP plane. The strong anisotropy of BP electrons in its plane is due to the fact that effective number of carriers along the zigzag (ZZ) direction are 10 times greater than those along the armchair (AC) direction [147, 148]. The anisotropy of BP mainly affects its band structure, mechanical, electrical, thermal, and optical properties [138, 141, 149] along with the ZZ and AC direction.

Exfoliated monolayer or few-layer BP presents great potential in flexible thermoelectrics. It has high electron or hole mobility, which is ~220 and ~350 $cm^2 V^{-1} s^{-1}$, respectively [150, 151]. BP displays high isotropy of S, σ, and κ in ZZ and AC direction due to the isotropy structures [152, 153]. The TE properties of BP nanoribbons have been studied by using the suspended micro-bridge method [154]. The S and σ of BP are ~270–320 $\mu V\ K^{-1}$ and ~700–1200 $S\ m^{-1}$, respectively. With the assistance of the first-principle calculation, it has been observed that thermoelectric properties of BP can be easily tuned, which gives insights to the thermoelectric transport of BP. The calculated S of BP monolayer is ~450 $\mu V\ K^{-1}$ at 300 K [155].

Compared to other 2D materials such as graphene (i.e. ~1800 $W\ m^{-1}\ K^{-1}$), the κ of BP is almost one or two orders of magnitude lower. For instance, as reported by McGaughey [156], the κ of monolayer BP was measured to be 110 and 36 $W\ m^{-1}\ K^{-1}$ in the ZZ and AC directions, respectively. Luo et al. applied the confocal micro-Raman spectroscopy method and measured the κ of few-layer BP [157], which is ~10–20 $W\ m^{-1}\ K^{-1}$ along AC direction and ~20–40 $W\ m^{-1}\ K^{-1}$ along the ZZ direction when the thickness increases from 9.5 to 15 nm, indicating strong thickness dependence and significant anisotropy in the in-plane direction.

Similar to other semiconductors, there are some common strategies to improve the thermoelectric properties of BP. For instance, alloying, doping, or compositing in BP are necessary to optimize the carrier concentration to balance the σ, S, and κ for high thermoelectric figure of merit. Bilu Liu et al. added arsenic into BP with heating [158] through alloying method thus resulting in a tunable band gap range of 0.15–0.3 eV. The n-type or p-type doping can be achieved by replacing phosphorus atoms with other elements, such as Cu, K, Te, As, Se, S, N, B, etc. [159]. For example, with increasing K coverage from 0.06 to 0.18 ml, VB was shifted down rapidly and the corresponding bandgap of phosphorene was gradually decreased [160]. The materials hybridized with BP may include carbon materials, metals, semiconductors, etc. The hybridization of BP with dissimilar components offers the advantages to create novel functional nanostructures with integrated properties [159].

More interestingly, the thermoelectric performance of BP is very sensitive to force which is different from other thermoelectric materials. This phenomenon can be explained by the different energy gap changes under compress and tensile. Specifically the compressive strain may suppress the band gap, and ultimately close the gap, while the tensile strain may enlarge the band gap [161]. For example, a compressive strain of 1–2% enhances the thermoelectric performance of the n-type doped BP, while tensile strain benefits the p-type doped BP [161–163].

Although the ZT value of BP is often below 1 [164], the first-principles simulations of the single-layer BP predict a ZT value greater than 1 at room temperature and even probably up to 2.5 at 500 K [165]. The ZT value of BP nanoribbons can be as high as 6.4 at room temperature when it has a 9-APNR structure (APNR is armchair phosphorene nanoribbons, 9 is the number of dimer lines across the ribbon width) [166].

7.4.3 Mxenes

MXene is a novel type of conducting layered materials [167]. MXenes are produced by selective etching of A element from the MAX phases, which are typically composed of layered metallic solids with chemical formula of $M_{n+1}AX_n$, where "M" is an early transition metal, A is III or IV element, and "X" is carbon or nitrogen [168]. The covalent bonds of M–X in MAX are much stronger than that of M–A metallic bond, therefore the A atom layer can be easily etched off with high-concentration hydrofluoric acid (HF), leaving the M–X atom layers forming layered MXene atomic crystals [169–171] (Figure 7.10a).

Most MXenes have high σ (~10^4 S cm^{-1} for the most well-studied $Ti_3C_2T_x$) [172, 173]. The exfoliation process of MAX phases and formation of MXenes have been shown in Figure 7.10b [172]. The M-A metal bond is destroyed by the stripping of atomic layer A and atoms in two-dimensional materials are firmly imprisoned in the layer by covalent bonds. The layers are relatively independent from each other and connected by the weak van der Waals force. Thus atoms only move in the layer without running out of the layer. The relatively limited range of motion makes electrons move very fast, which greatly improves MXenes' carrier mobility [170]. The layer structures of MXenes have been shown in Figure 7.10c–f [172]. The interlayer spacing of MXenes also can be tuned by changing the intercalated species between nanosheets. The surface of MXenes can be functionalized with various chemical groups. In addition, the large surface area inherited from their 2D structure makes MXenes promising in electrochemical energy storage applications. These properties make MXenes unique among all known 2D materials.

Recently, MXenes have been theoretically and experimental investigated for their potential application in flexible thermoelectrics. A few MXenes, including Sc_2CT_x, Zr_2CT_x, Ti_2CT_x, and Hf_2CT_x, are predicted to be

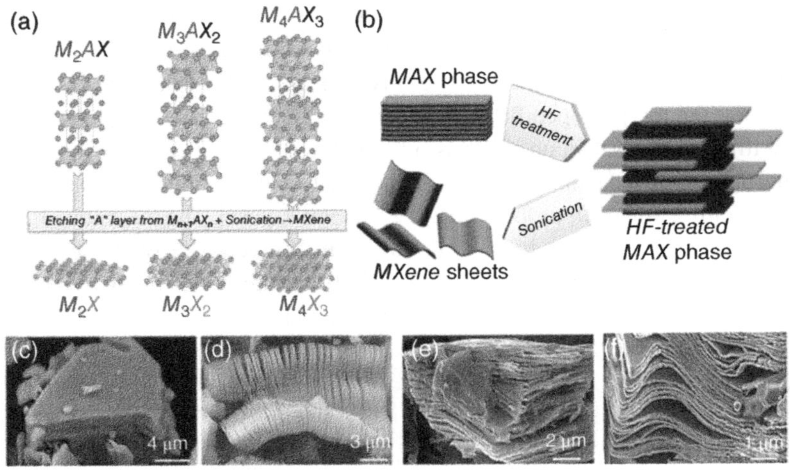

Figure 7.10 (a) Structure of MAX phases and the corresponding MXenes. *Source:* Naguib et al. [171]. © John Wiley & Sons. (b) Schematic illustration of the exfoliation process of MAX phases and formation of MXenes [172]. (c–f) Secondary electron SEM micrographs for (c) Ti_3AlC_2 particle before treatment, which is typical of unreacted MAX phases, (d) Ti_3AlC_2 after HF treatment, (e) Ta_4AlC_3 after HF treatment, and (f) TiNbAlC after HF treatment [172]. *Source:* Naguib et al. [172]. © American Chemical Society.

semiconducting with nonzero band gaps [174]. First principles calculations have revealed that these MXenes possess high S. For example, the thermoelectric properties of Ti_2CO_2, Hf_2CO_2, Zr_2CO_2, Sc_2CF, $Sc_2C(OH)_2$, and Sc_2CO_2 were simulated and it was found that they possessed the high S at 100 K [175], e.g. 1140 $\mu V K^{-1}$ for Ti_2CO_2 and 2200 $\mu V K^{-1}$ for $Sc_2C(OH)_2$ at 100 K [176].

The S, σ, and PF of monolayers and multilayers of the functionalized MXenes are investigated with first-principles calculation and Boltzmann transport calculation methods [177]. The monolayered and multilayered M_2C (M = Sc, Ti, V, Zr, Nb, Mo, Hf, and Ta), M_2N (M = Ti, Zr, and Hf), and MXenes functionalized with F, OH, and O groups were explored. In calculations, the multilayer systems are simulated by three-dimensional close stack of the functionalized monolayers. It turns out that in the electronic and thermoelectric behaviors, no significant differences between the monolayers and their corresponding multilayer structures are observed.

Figure 7.11a shows the DOS, carrier density, S, σ/τ, and $S^2\sigma/\tau$ of multilayer Mo_2CF_2, $Sc_2C(OH)_2$, and Nb_2CF_2, as a function of the chemical

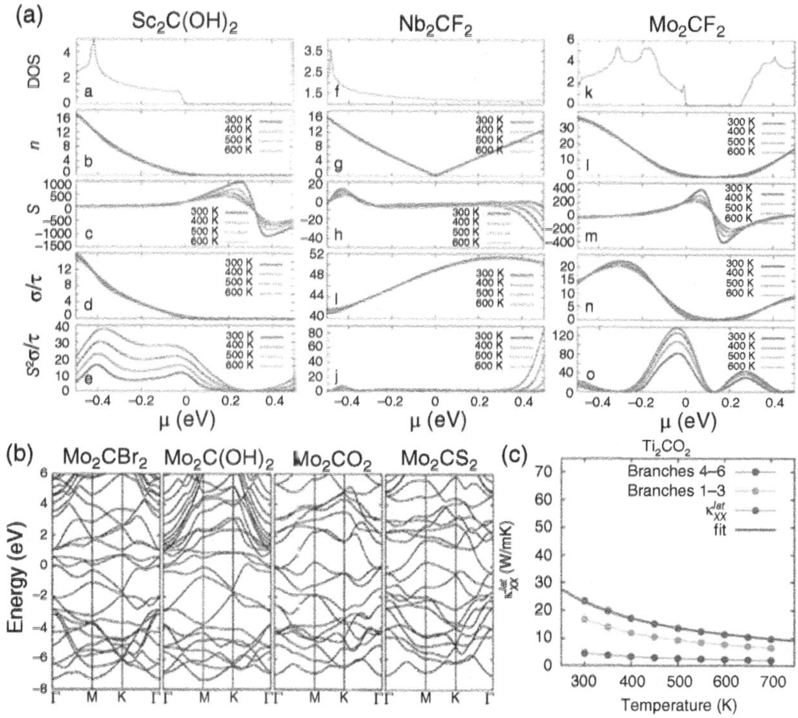

Figure 7.11 (a) DOS [states/eV/cell], carrier density (n [10^{20} cm^3]) and the S (S [μV K^{-1}]; $S = (Sxx + Syy + Szz)/3$), σ (σ/τ [10^{17} Ω^{-1}cm^{-1}s^{-1}]; $\sigma = (\sigma_{xx} + \sigma_{yy} + \sigma_{zz})/3$), and the power factor ($S^2\sigma/\tau$ [10^4 μW cm^{-1}K^{-2}s^{-1}]) as a function of the chemical potential at various temperatures for multilayers of $Sc_2C(OH)_2$ (a–e), Nb_2CF_2 (f–j) and Mo_2CF_2 (k–o). $\mu = 0$ is set to be at the original highest occupied level. It is noted that both valence and conduction bands have been included in the transport calculations, but the results were shown only for the range of chemical potentials between −0.5 and 0.5 eV. (b) Typical band structures of semiconducting MXenes. *Source:* Khazaei et al. [176]. © Royal Society of Chemistry. (c) Lattice contributions to the thermal conductivity as functions of the temperature. *Source:* Gandi et al. [178]. © American Chemical Society.

potential and temperature in the range of 300–600 K [179]. From calculations, it is found that all semiconducting MXenes, e.g. Sc_2CF_2, $Sc_2C(OH)_2$, Ti_2CO_2, Zr_2CO_2, Mo_2CF_2, Mo_2CCl_2, $Mo_2C(OH)_2$, and Hf_2CO_2 have large S (4100 μV K^{-1} at 400 K) near the band edges. The effect of band gap on the enhancement of S can be revealed by the formula in which the S is proportional to ν/σ, where σ is the electrical conductivity and ν is the $(\varepsilon - \mu)$ weighted average value of squared band velocities of electrons multiplied by the DOS and the derivative of the Fermi function $(f(\varepsilon))$, with respect to the energy $(df(\varepsilon)/d(\varepsilon))$. The $(\varepsilon - \mu)df(\varepsilon)/d(\varepsilon)$ has opposite signs in energy regions above and below the chemical potential [180]. In semiconducting materials, as the chemical potential moves toward band edges, the difference between the number of states above and below the chemical potential increases, which results in the increase in S. Due to this reason, the semiconducting MXenes such as $Sc_2C(OH)_2$ and Mo_2CF_2 possess large S near the band edges. The S values of semiconducting MXenes are sensitive to the temperature in a way that S decreases as the temperature increases from 300 to 600 K.

Although all semiconducting MXenes demonstrate large S (>100 μV K^{-1} at 400 K), only some of them hold high σ near the band edges. The fundamental reason for the superior σ of semiconducting Mo_2CF_2, $Mo_2C(OH)_2$, and Mo_2CCl_2 than other semiconducting MXenes, Sc_2CF_2, $Sc_2C(OH)_2$, Ti_2CO_2, Zr_2CO_2, and Hf_2CO_2 dates back to the difference in their band structures. For instance, the peculiar band structure of Mo_2CF_2 resulted in higher σ than that of other semiconductor, as shown in Figure 7.11b [176, 178]. In this structure, the curves below the valence band maximum (t_{2g}) are relatively flat (Γ to M points and K to Γ) and other portion of the curves are linearly dispersive (M to K). Hence, when the E_F locates near the valence band maximum of such a band structure, a relatively large Fermi surface is constructed, which holds a relatively large Fermi velocity. Consequently, Mo_2CF_2 shows a higher σ than that of Ti_2CO_2 and $Sc_2C(OH)_2$.

The κ of various MXenes, such as M_2CO_2 (M = Ti, Zr, Hf), Sc_2CT_2 (T = F, O, OH), and Mo_2C, has also been studied in consideration of both phonon and electron contributions [179, 181–184]. At room temperature, the κ of Hf_2CO_2 is 62.12–131.2 (27.63–53.03) W m^{-1} K^{-1} along the armchair (zigzag) flake length of 1–100 μm [179]. Among Sc_2CT_2 (T = F, O, OH), $Sc_2C(OH)_2$ is predicted to be an excellent candidate for thermoelectric applications at elevated temperatures higher than 300 K due to its relatively good S, high σ, and very low lattice κ [183]. Since the PF is high, the ZT can be strongly enhanced by lowering the lattice thermal conductivity with O and F functionalization. In fact, the lowest lattice κ can be achieved by the OH functionalization, due to the largest anharmonic phonon scattering. The κ of graphene (10-μm sample), blue phosphorene, MoS_2 (10-μm sample), $MoSe_2$ (1-μm sample), and WSe_2 (1-μm sample) at room temperature are given for comparison, as shown in Table 7.2 [185].

Table 7.2 Lattice thermal conductivity and contribution of the acoustic phonon branches ($W m^{-1} K^{-1}$). Data for graphene (10-µm sample), blue phosphorene, MoS_2 (10-µm sample), $MoSe_2$ (1-µm sample), and WSe_2 (1-µm sample) at room temperature are given for comparison [185].

Samples	κ_l	$\kappa_l^{acoustic}$
Sc_2CO_2	59	48
Sc_2CF_2	36	30
$ScC(OH)_2$	10	8
Graphene	3600	3564
Blue phosphorene	78	76
MoS_2	108	98
$MoSe_2$	69	69
WSe_2	43	42

Source: Ref. [185]. © American Physical Society.

References

1 Nazmitdinov, R.G. (2009). *Phys. Part. Nucl.* 40: 71.
2 Li, L. and Xin, Y. (2010). *J. Phys. Chem. Lett.* 1: 2572.
3 Wang, R.Y., Feser, J.P., Lee, J.S. et al. (2008). *Nano Lett.* 8: 2283.
4 Dajiang, Y., Chenguang, L., Huiming, Y., and Herman, I.P. (2013). *Nanoscale* 5: 7290.
5 Khitun, A., Balandin, A., Liu, J., and Wang, K. (2000). *J. Appl. Phys.* 88: 696.
6 Liu, J., Khitun, A., Wang, K. et al. (2001). *J. Crystal Growth* 227: 1111.
7 Hassaine, M., Ramos, M.A., Krivchikov, A.I. et al. (2012). *Phys. Rev. B Condens. Matter* 85: 131[illegible].
8 Chen, L., Wang, X., and Kumar, S. (2015). *Sci. Rep.* 5: 12763.
9 Almutlaq, N., Algaliby, Q., Bailey, S., and Lambert, C.J. (2016). *Nanoscale* 8: 13597.
10 Yan, Y., Liang, Q.F., Zhao, H., and Wu, C.Q. (2012). *Phys. Lett. A* 376: 1154.
11 Mao, S., Harada, K., Ikeda, M. et al. (2011). *Appl. Phys. Lett.* 99: 188.
12 Efimov, V., Mezhov-Deglin, L., and Nikolaev, R. (1997). *J. Exp. Theor. Phys. Lett.* 65: 687.
13 Yu, R., Tea, N., Salamon, M. et al. (1992). *Phys. Rev. Lett.* 68: 2050.
14 Lambert, C. (2015). *Chem. Soc. Rev.* 44: 875.
15 Li, S., Fan, T., Liu, X. et al. (2017). *ACS Appl. Mater. Interfaces* **9**: 3677.
16 Zhu, Y., Murali, S., Cai, W. et al. (2010). *Adv. Mater.* **22**: 3906.

17 Hong, M., Chen, Z.-G., Yang, L., and Zou, J. (2016). *Nanoscale* 8: 8681.

18 Murphy, C.J. and Jana, N.R. (2002). *Adv. Mater.* 14: 80.

19 Liu, R. and Lee, S.B. (2008). *J. Am. Chem. Soc.* 130: 2942.

20 Joo, J., Kim, B.H., Park, D.H. et al. (2005). *Synth. Met.* 153: 313.

21 Yusupov, K., Stumpf, S., You, S.J. et al. (2018). *Adv. Funct. Mater.* 28: 11.

22 Martin, J., Campoy-Quiles, M., Nogales, A. et al. (2014). *Soft Matter* 10: 3335.

23 Liu, H.Q., Kameoka, J., Czaplewski, D.A., and Craighead, H.G. (2004). *Nano Lett.* 4: 671.

24 Chronakis, I.S., Grapenson, S., and Jakob, A. (2006). *Polymer* 47: 1597.

25 Laforgue, A. and Robitaille, L. (2010). *Macromolecules* 43: 4194.

26 Zhang, J., Zhang, K., Xu, F. et al. (2018). *Compos. Part B* 136: 234.

27 Kim, M.S., Kim, B.G., and Kim, J. (2009). *ACS Appl. Mater. Interfaces* 1: 1264.

28 Kim, J.S., Lee, J.H., Park, J.H. et al. (2011). *Adv. Funct. Mater.* 21: 480.

29 Lim, K.H., Wong, K.W., Cadavid, D. et al. (2019). *Compos. Part B* 164: 54.

30 Taggart, D.K., Yang, Y., Kung, S.C. et al. (2011). *Nano Lett.* 11: 125.

31 Cho, B., Park, K.S., Baek, J. et al. (2014). *Nano Lett.* 14: 3321.

32 Wang, Y., Cai, K., and Yao, X. (2011). *ACS Appl. Mater. Interfaces* 3: 1163.

33 Wu, J., Sun, Y., Xu, W., and Zhang, Q. (2014). *Synth. Met.* 189: 177.

34 Liang, L., Chen, G., and Guo, C.-Y. (2017). *Mater. Chem. Front.* 1: 380.

35 Du, Y., Niu, H., Li, J. et al. (2018). *Polymers (Basel)*: 10.

36 Rojo, M.M., Martin, J., Grauby, S. et al. (2014). *Nanoscale* 6: 7858.

37 Lu, G., Bu, L., Li, S., and Yang, X. (2014). *Adv. Mater.* 26: 2359.

38 Wang, X., Ho, V., Segalman, R.A., and Cahill, D.G. (2013). *Macromolecules* 46: 4937.

39 Groenendaal, B.L., Jonas, F., Freitag, D. et al. (2000). *Adv. Mater.* 12: 481.

40 Boyle, C.J., Upadhyaya, M., Wang, P. et al. (2019). *Nat. Commun.* 10: 2827.

41 Chen, X., Shi, W., and Zhang, K. (2020). *ACS Appl. Mater. Interfaces* 12: 34451.

42 Shirai, Y., Takami, S., Lasmono, S. et al. (2011). *J. Polym. Sci. B Polym. Phys.* 49: 1762.

43 Wang, Y., Cai, K., Chen, S. et al. (2014). *J. Nanopart. Res.*: 16.

44 Huang, J.X. and Kaner, R.B. (2004). *J. Am. Chem. Soc.* 126: 851.

45 Han, M.G. and Foulger, S.H. (2006). *Small* 2: 1164.

46 Xue, Y., Gao, C., Liang, L. et al. (2018). *J. Mater. Chem. A* 6: 22381.

47 Stejskal, J. and Gilbert, R.G. (2002). *Pure Appl. Chem.* 74: 857.

48 Yan, H., Ohno, N., and Toshima, N. (2000). *Chem. Lett.*: 392.

49 Sun, Y., Wei, Z., Xu, W., and Zhu, D. (2010). *Synth. Met.* 160: 2371.

50 Liu, Y., Zhang, H., Porwal, H. et al. (2019). *Polym. Int.* 68: 299.

51 Wu, Q.F., He, K.X., Mi, H.Y., and Zhang, X.G. (2007). *Mater. Chem. Phys.* 101: 367.

52 Merlo, J.A. and Frisbie, C.D. (2003). *J. Polym. Sci.B Polym. Phys.* 41: 2674.

53 Yang, X., Loos, J., Veenstra, S.C. et al. (2005). *Nano Lett.* 5: 579.
54 Sun, J., Yeh, M.L., Jung, B.J. et al. (2010). *Macromolecules* 43: 2897.
55 He, M., Ge, J., Lin, Z. et al. (2012). *Energy Environ. Sci.*: 5.
56 Yue, R. and Xu, J. (2012). *Synth. Met.* 162: 912.
57 Liu, J.H., Arif, M., Zou, J.H. et al. (2009). *Macromolecules* 42: 9390.
58 Wang, Y., Zhou, J., and Yang, R. (2011). *J. Phys. Chem. C* 115: 24418.
59 Lu, G.H., Tang, H.W., Qu, Y.P. et al. (2007). *Macromolecules* 40: 6579.
60 Canetta, C., Guo, S., and Narayanaswamy, A. (2014). *Rev. Sci. Instrum.* 85: 104901.
61 Iijima, S. (1991). *Nature* 354: 56.
62 Terrones, M. (2003). *Annu. Rev. Mater. Res.* 33: 419.
63 Mittal, G., Dhand, V., Rhee, K.Y. et al. (2015). *J. Ind. Eng. Chem.* 21: 11.
64 Venema, L.C., Janssen, J.W., Buitelaar, M.R. et al. (2000). *Phys. Rev. B* 62: 5238.
65 Jorio, A., Saito, R., Hafner, J.H. et al. (2001). *Phys. Rev. Lett.* 86: 1118.
66 Wilder, J.W.G., Venema, L.C., Rinzler, A.G. et al. (1998). *Nature* 391: 59.
67 Odom, T.W., Huang, J.-L., Kim, P., and Lieber, C.M. (1998). *Nature* 391: 62.
68 O'Connell, M.J., Bachilo, S.M., Huffman, C.B. et al. (2002). *Science* 297: 593.
69 Saito, R., Fujita, M., Dresselhaus, G., and Dresselhaus, M.S. (1992). *Appl. Phys. Lett.* 60: 2204.
70 Ebbesen, T.W., Lezec, H.J., Hiura, H. et al. (1996). *Nature* 382: 54.
71 Carroll, D.L., Redlich, P., Ajayan, P.M. et al. (1997). *Phys. Rev. Lett.* 78: 2811.
72 Gao, R.P., Pan, Z.W., and Wang, Z.L. (2001). *Appl. Phys. Lett.* 78: 1757.
73 Derycke, V., Martel, R., Appenzeller, J., and Avouris, P. (2002). *Appl. Phys. Lett.* 80: 2773.
74 Yarali, M., Hao, J., Khodadadi, M. et al. (2017). *RSC Adv.* 7: 14078.
75 Collins, P.G., Bradley, K., Ishigami, M., and Zettl, A. (2000). *Science* 287: 1801.
76 Tans, S.J., Verschueren, A.R.M., and Dekker, C. (1998). *Nature* 393: 49.
77 Javey, A., Guo, J., Wang, Q. et al. (2003). *Nature* 424: 654.
78 Peng, L.M., Zhang, Z., and Wang, S. (2014). *Mater. Today* 17: 433.
79 Kong, J., Zhou, C., Morpurgo, A. et al. (1999). *Appl. Phys. A* 69: 305.
80 Bezryadin, A., Verschueren, A.R.M., Tans, S.J., and Dekker, C. (1998). *Phys. Rev. Lett.* 80: 4036.
81 Langer, L., Bayot, V., Grivei, E. et al. (1996). *Phys. Rev. Lett.* 76: 479.
82 Fujii, M., Zhang, X., Xie, H. et al. (2005). *Phys. Rev. Lett.* 95: 065502.
83 Lezec, H.H.H.J., Bennett, J.W., Ebbesen, T.W. et al. (1996). *Nat. Mater.* 382: 55.
84 Shi, L., Li, D., Yu, C. et al. (2003). *J. Heat Transf.* 125: 881.
85 Kim, P., Shi, L., Majumdar, A., and McEuen, P.L. (2001). *Phys. Rev. Lett.* 87: 215502.

86 Yu, C., Shi, L., Yao, Z. et al. (2005). *Nano Lett.* 5: 1842.
87 Pop, E., Mann, D., Wang, Q. et al. (2006). *Nano Lett.* 6: 96.
88 Choi, T.Y., Poulikakos, D., Tharian, J., and Sennhauser, U. (2006). *Nano Lett.* 6: 1589.
89 Choi, T.Y., Poulikakos, D., Tharian, J., and Sennhauser, U. (2005). *Appl. Phys. Lett.*: 87.
90 Lu, L., Yi, W., and Zhang, D.L. (2001). *Rev. Sci. Instrum.* 72: 2996.
91 Wang, L., Zhang, J., Guo, Y. et al. (2019). *Carbon* 148: 290.
92 Wang, J., Cai, K., Shen, S., and Yin, J. (2014). *Synth. Met.* 195: 132.
93 Zhang, Q., Wang, W., Li, J. et al. (2013). *J. Mater. Chem. A* 1: 12109.
94 Du, Y., Shen, S.Z., Yang, W.D. et al. (2012). *Synth. Met.* 162: 375.
95 Ren, F., Wang, H., Menchhofer, P.A., and Kiggans, J.O. (2013). *Appl. Phys. Lett.*: 103.
96 Geim, A.K. and Novoselov, K.S. (2007). *Nat. Mater.* 6: 183.
97 Geim, A.K. (2009). *Science* 324: 1530.
98 Ren, X., Lian, P., Xie, D. et al. (2017). *J. Mater. Sci.* 52: 10364.
99 Zhang, X., Hsu, C.-H., Ren, X. et al. (2015). *Angew. Chem. Int. Ed.* 54: 114.
100 Seiki, N., Shoji, Y., Kajitani, T. et al. (2015). *Science* 348: 1122.
101 Leung, F.K.C., Ishiwari, F., Kajitani, T. et al. (2016). *J. Am. Chem. Soc.* 138: 11727.
102 He, D., Zhang, Y., Wu, Q. et al. (2014). *Nat. Commun.* 5: 5162.
103 Xu, C., He, P., Liu, J. et al. (2016). *Angew. Chem.* 55: 9519.
104 Kissel, P., Erni, R., Schweizer, W.B. et al. (2012). *Nat. Chem.* 4: 287.
105 Champness, N.R. (2014). *Nat. Chem.* 6: 757.
106 Kissel, P., Murray, D.J., Wulftange, W.J. et al. (2014). *Nat. Chem.* 6: 774.
107 Kory, M.J., Wörle, M., Weber, T. et al. (2014). *Nat. Chem.* 6: 779.
108 Yang, F., Cheng, S., Zhang, X. et al. (2018). *Adv. Mater.*: 30.
109 Sedlovets, D.M., Shuvalov, M.V., Vishnevskiy, Y.V. et al. (2013). *Mater. Res. Bull.* 48: 3955.
110 Jiang, L., Niu, T., Lu, X. et al. (2013). *J. Am. Chem. Soc.* 135: 9050.
111 Zhang, J., Li, J., Wang, Z. et al. (2014). *Chem. Mater.* 26: 2460.
112 Dresselhaus, M.S., Chen, G., Tang, M.Y. et al. (2007). *Adv. Mater.* 19: 1043.
113 Koumoto, K., Wang, Y., Zhang, R. et al. (2010). *Annual Review of Materials Research*, vol. 40 (eds. D.R. Clarke, M. Ruhle and F. Zok), 363.
114 Novoselov, K.S., Geim, A.K., Morozov, S.V. et al. (2004). *Science* 306: 666.
115 Allen, M.J., Tung, V.C., and Kaner, R.B. (2010). *Chem. Rev.* 110: 132.
116 Novoselov, K.S., Morozov, S.V., Mohinddin, T.M.G. et al. (2007). *Phys. Status Solidi B* **244**: 4106.
117 Bolotin, K.I., Sikes, K.J., Jiang, Z. et al. (2008). *Solid State Commun.* 146: 351.
118 Kim, K.S., Zhao, Y., Jang, H. et al. (2009). *Nature* 457: 706.
119 Pop, E., Varshney, V., and Roy, A.K. (2012). *MRS Bull.* 37: 1273.

120 Dollfus, P., Viet Hung, N., and Saint-Martin, J. (2015). *J. Phys. Condens. Matter*: 27.

121 Zuev, Y.M., Chang, W., and Kim, P. (2009). *Phys. Rev. Lett.*: 102.

122 Wei, P., Bao, W., Pu, Y. et al. (2009). *Phys. Rev. Lett.*: 102.

123 Park, J., He, G., Feenstra, R.M., and Li, A.P. (2013). *Nano Lett.* 13: 3269.

124 Xu, X., Pereira, L.F.C., Wang, Y. et al. (2014). *Nat. Commun.* 5: 3689.

125 Balandin, A.A., Ghosh S., Bao, W. et al. (2008). *Nano Lett.* 8: 902.

126 Wu, Z.S., Ren, W., Gao, L. et al. (2009). *Carbon* 47: 493.

127 Li, X., Yin, J., Zhou, J. et al. (2012). *Appl. Phys. Lett.*: 100.

128 Okuda, M. and Ohkubo, S. (1992). *Thin Solid Films* 213: 176.

129 Xu, X., Pereira, L.F.C., Wang, Y. et al. (2014). *Nat. Commun.*: 5.

130 Calizo, I., Balandin, A.A., Bao, W. et al. (2007). *Nano Lett.* 7: 2645.

131 Lee, J.U., Yoon, D., Kim, H. et al. (2011). *Phys. Rev. B* 83: 081419.

132 Calizo, I., Bao, W., Miao, F. et al. (2007). *Appl. Phys. Lett.* 91: 201904.

133 Korotcenkov, G. (2019). *Sensors (Basel)*: 19.

134 Lin, S., Li, Y., Qian, J., and Lau, S.P. (2019). *Mater. Today Energy* 12: 1.

135 Ling, X., Wang, H., Huang, S. et al. (2015). *Proc. Natl. Acad. Sci. U S A* 112: 4523.

136 Cai, Y.Q., Zhang, G., and Zhang, Y.W. (2014). *Sci. Rep.* 4: 6.

137 Zhou, F., Ouyang, L., Zeng, M. et al. (2019). *J. Alloys Compd.* 784: 339.

138 Anju, S., Ashtami, J., and Mohanan, P.V. (2019). *Mater. Sci. Eng. C Mater. Biol. Appl.* 97: 978.

139 Wu, S., Hui, K.S., and Hui, K.N. (2018). *Adv. Sci. (Weinh)* 5: 1700491.

140 Eswaraiah, V., Zeng, Q., Long, Y., and Liu, Z. (2016). *Small* 12: 3480.

141 Mu, X., Wang, J., and Sun, M. (2019). *Mater. Today Phys.* 8: 92.

142 Zhang, C.D., Lian, J.C., Yi, W. et al. (2009). *J. Phys. Chem. C* 113: 18823.

143 Wang, V., Kawazoe, Y., and Geng, W.T. (2015). *Phys. Rev. B*: 91.

144 Tran, V., Soklaski, R., Liang, Y., and Yang, L. (2014). *Phys. Rev. B* 89: 235319.

145 Castellanos-Gomez, A. (2015). *J. Phys. Chem. Lett.* 6: 4280.

146 Liu, H., Neal, A.T., Zhu, Z. et al. (2014). *ACS Nano* 8: 4033.

147 Kumar, P., Bhadoria, B.S., Kumar, S. et al. (2016). *Phys. Rev. B* 93: 195428.

148 Xia, F., Wang, H., and Jia, Y. (2014). *Nat. Commun.* 5: 4458.

149 Xing, C., Zhang, J., Jing, J. et al. (2019). *Chem. Eng. J.* 370: 120.

150 Appalakondaiah, S., Vaitheeswaran, G., Lebègue, S. et al. (2012). *Phys. Rev. B* 86: 035105.

151 Sun, J., Zheng, G., Lee, H.-W. et al. (2014). *Nano Lett.* 14: 4573.

152 Zhang, Y., Zheng, Y., Rui, K. et al. (2017). *Small*: 13.

153 Wang, Y., Xu, G., Hou, Z. et al. (2016). *Appl. Phys. Lett.*: 108.

154 Lee, S., Yang, F., Suh, J. et al. (2015). *Nat. Commun.* 6: 8573.

155 Liao, B., Zhou, J., Qiu, B. et al. (2015). *Phys. Rev. B* 91: 235419.

156 Jain, A. and McGaughey, A.J.H. (2015). *Sci. Rep.* 5: 8501.

157 Luo, Z., Maassen, J., Deng, Y. et al. (2015). *Nat. Commun.* 6: 8572.

158 Liu, B., Köpf, M., Abbas, A.N. et al. (2015). *Adv. Mater.* 27: 4423.

159 Lei, W., Liu, G., Zhang, J., and Liu, M. (2017). *Chem. Soc. Rev.* 46: 3492.

160 Kim, J., Baik, S.S., Ryu, S.H. et al. (2015). *Science* 349: 723.

161 Qin, G., Yan, Q.B., Qin, Z. et al. (2014). *Sci. Rep.* 4: 6946.

162 Rodin, A.S., Carvalho, A., and Castro Neto, A.H. (2014). *Phys. Rev. Lett.* 112: 176801.

163 Lv, H.Y., Lu, W.J., Shao, D.F., and Sun, Y.P. (2014). *Phys. Rev. B* 90 (8): 085433.

164 Duan, S., Cui, Y., Chen, X. et al. (2019). *Adv. Funct. Mater.* 29 (38): 1904346.

165 Fei, R., Faghaninia, A., Soklaski, R. et al. (2014). *Nano Lett.* 14: 6393.

166 Zhang, J., Liu, H.J., Cheng, L. et al. (2014). *Sci. Rep.* 4: 6452.

167 Khazaei, M., Ranjbar, A., Arai, M. et al. (2017). *J. Mater. Chem. C* 5: 2488.

168 Chhowalla, M., Shin, H.S., Eda, G. et al. (2013). *Nat. Chem.* 5: 263.

169 Xu, C., Wang, L., Liu, Z. et al. (2015). *Nat. Mater.* 14: 1135.

170 Zha, X.H., Zhou, J., Zhou, Y. et al. (2016). *Nanoscale* 8: 6110.

171 Naguib, M., Mochalin, V.N., Barsoum, M.W., and Gogotsi, Y. (2014). *Adv. Mater.* 26: 992.

172 Naguib, M., Mashtalir, O., Carle, J. et al. (2012). *ACS Nano* 6: 1322.

173 Wang, Z., Kim, H., and Alshareef, H.N. (2018). *Adv. Mater.* 30: 1706656.

174 Pang, J., Mendes, R.G., Bachmatiuk, A. et al. (2019). *Chem. Soc. Rev.* 48: 72.

175 Khazaei, M., Arai, M., Sasaki, T. et al. (2013). *Adv. Funct. Mater.* 23: 2185.

176 Khazaei, M., Arai, M., Sasaki, T. et al. (2014). *Phys. Chem. Chem. Phys.* 16: 7841.

177 May, A.F., Singh, D.J., and Snyder, G.J. (2009). *Phys. Rev. B*: 79.

178 Gandi, A.N., Alshareef, H.N., and Schwingenschlögl, U. (2016). *Chem. Mater.* 28: 1647.

179 Hope, M.A., Forse, A.C., Griffith, K.J. et al. (2016). *Phys. Chem. Chem. Phys.* 18: 5099.

180 Yang, J., Li, H., Wu, T. et al. (2008). *Adv. Funct. Mater.* 18: 2880.

181 Hasan, M.Z. and Kane, C.L. (2010). *Rev. Mod. Phys.* 82: 3045.

182 He, J., Lyu, P., Sun, L.Z. et al. (2016). *J. Mater. Chem. C*: 4.

183 Heyd, J., Scuseria, G.E., and Ernzerhof, M. (2006). *J. Chem. Phys.* 124 (21): 219906.

184 Hong, H., O'Keeffe, K.P., and Strogatz, S.H. (2016). *Phys. Rev. E* 93 (2): 022219.

185 Kumar, S. and Schwingenschlögl, U. (2016). *Phys. Rev. B* 94 (33): 035405.

Index

Printed and bound by CPI Group (UK) Ltd, Croydon, CR0 4YY
17/09/2026
14964318-0001